A ROADMAP FOR FORMAL PROPERTY VERIFICATION

A Roadmap for Formal Property Verification

by

PALLAB DASGUPTA
Indian Institute of Technology, Kharagpur, India

 Springer

A C.I.P. Catalogue record for this book is available from the Library of Congress.

ISBN-10 1-4020-4757-6 (HB)
ISBN-13 978-1-4020-4757-2 (HB)
ISBN-10 1-4020-4758-4 (e-book)
ISBN-13 978-1-4020-4758-9 (e-book)

Published by Springer,
P.O. Box 17, 3300 AA Dordrecht, The Netherlands.

www.springer.com

Printed on acid-free paper

To Sagarika ... for her understanding and support,
Parnika ... for her invasions,
and my dear parents

Contents

Preface xi

1 Introduction 1
   1.1 Writing Our First Formal Specification 2
   1.2 Is My Specification Correct? 5
   1.3 Have I written Enough Properties? 6
   1.4 Property Verification 8
        1.4.1 The Dynamic ABV Approach 9
        1.4.2 The FPV Approach 11
   1.5 Verification by Specification Refinement 14
   1.6 The New Flow 17

2 Languages for Temporal Properties 19
   2.1 The Basic Temporal Operators 22
        2.1.1 Intuitive Explanation 23
        2.1.2 Formal Semantics 24
   2.2 Logics for Temporal Specification 25
        2.2.1 Linear Temporal Logic 26
        2.2.2 Computation Tree Logic 27
        2.2.3 LTL versus CTL 29
        2.2.4 Real Time Temporal Logics 30
   2.3 SystemVerilog Assertions 31
        2.3.1 A Quick Overview 32
        2.3.2 The Notion of Sequences 37
        2.3.3 Sequence Operations: Repetition 38
        2.3.4 Sequence Operations: AND, INTERSECT, OR 40
        2.3.5 Local Variables 41
        2.3.6 Properties 42
        2.3.7 Assume-Assert Specifications 44
        2.3.8 The Cover Statement 47
   2.4 Architectural Styles for Assertion IPs 47

    2.4.1   The Main Steps   49
    2.4.2   Coding Styles: *Event and State-based Checkers*   50
  2.5   Concluding Remarks   56
  2.6   Bibliographic Notes   57

3   How Does the Property Checker Work?   59
  3.1   Checkers are State Machines!   61
  3.2   The Verification Strategy   63
  3.3   Dynamic Property Verification   64
  3.4   Formal Property Verification   67
    3.4.1   Creating the Checker Automaton   68
    3.4.2   FSM Extraction   72
    3.4.3   Computing the Product   74
  3.5   BDD-based Formal Property Verification   78
    3.5.1   What is a BDD?   79
    3.5.2   BDDs for State Machines   83
    3.5.3   Symbolic Reachability   84
    3.5.4   CTL Model Checking   90
  3.6   SAT-based Formal Property Verification   92
    3.6.1   What is SAT?   92
    3.6.2   SAT-based Reachability   93
    3.6.3   Detecting Loops   95
    3.6.4   Unfolding Properties   96
    3.6.5   Bounded Model Checking   96
  3.7   Concluding Remarks   97
  3.8   Bibliographic Notes   99

4   Is My Specification Consistent? 101
  4.1   Satisfiability and Vacuity 103
    4.1.1   Writing Unsatisfiable Specifications 103
    4.1.2   The Notion of Vacuity 105
  4.2   Satisfiability is not Enough 105
    4.2.1   Realizability 107
    4.2.2   Receptiveness 109
  4.3   Games with the Environment 114
  4.4   Methods for Consistency Checking 115
    4.4.1   Alternating Automata and LTL 117
    4.4.2   Satisfiability Checks 119
    4.4.3   Realizability Checks 121
    4.4.4   Receptiveness Checks 124
  4.5   The SpecChecker Tool 125
  4.6   Concluding Remarks 126
  4.7   Bibliographic Notes 128

5   Have I Written Enough Properties? 129
   5.1   Simulation Coverage Metrics 131
   5.2   Mutation-based FPV Coverage 132
      5.2.1   Falsity and Vacuity Coverage 134
      5.2.2   FSM Coverage 135
      5.2.3   Code and Circuit Coverage 136
   5.3   Structural Versus Functional Coverage 136
   5.4   Fault-based FPV Coverage 138
      5.4.1   The Coverage Strategy 141
      5.4.2   Coverage of Faults on Non-inputs 142
      5.4.3   Coverage of Faults on Inputs 145
      5.4.4   LTL-Covanalyzer: The Tool 148
      5.4.5   Building it Over FPV Tools 149
      5.4.6   Other Fault Models 153
   5.5   Concluding Remarks 154
   5.6   Bibliographic Notes 155

6   Design Intent Coverage 157
   6.1   An Introductory Example 160
      6.1.1   Priority Cache Access: *Architectural Specs* 160
      6.1.2   Is my Implementation Plan Correct? 162
      6.1.3   The Correct Architecture 165
      6.1.4   How Does Design Intent Coverage Help? 166
   6.2   The Formal Problem 167
      6.2.1   Where is the Coverage Gap? 171
      6.2.2   How Should We Present the Coverage Hole? 173
   6.3   The Intent Coverage Algorithm 174
      6.3.1   Unfolding the Coverage Gap 175
      6.3.2   Elimination of Non-architectural Signals 177
      6.3.3   The Term Distribution Algorithm 178
      6.3.4   Weakening Unbounded Temporal Properties 182
      6.3.5   Special Treatment of Invariant Properties 184
   6.4   Soundness of the Intent Coverage Algorithm 186
   6.5   Multi-property Representation of the Coverage Gap 186
   6.6   SpecMatcher – The Intent Coverage Tool 188
   6.7   Priority Cache Access – *A Closer Look* 189
   6.8   Concluding Remarks 192
   6.9   Bibliographic Notes 193

7   Test Generation Games 195
   7.1   Constrained Random Test Generation 196
      7.1.1   Layered Verification Methodologies 197
      7.1.2   The Benefits 199
      7.1.3   The Limitations 201
      7.1.4   Dynamic Coverage Driven Verification 202

7.2   Assertions Viewed as Coverage Points! 203
7.3   Games with the Environment 204
     7.3.1   Vacuity Games 205
     7.3.2   Realizability Games 207
7.4   Intelligent Test Generation for Property Coverage 208
     7.4.1   Dynamic Monitoring 208
     7.4.2   Online Test Generation 209
     7.4.3   Test Generation Algorithms 210
7.5   The Integrated Verification Flow 212
7.6   Concluding Remarks 214
7.7   Bibliographic Notes 215

8   A Roadmap for Formal Property Verification 217
8.1   Simulation-based Validation Flow 218
8.2   Formal Verification Flow 221
8.3   The Three Pillars 224
     8.3.1   What's Between Simulation and Model Checking? 229
     8.3.2   What's between Intent Coverage and Model Checking? . 231
     8.3.3   What's between Intent Coverage and Simulation? 233
8.4   The Integrated Flow 234
     8.4.1   Architectural Validation 235
     8.4.2   RTL Validation 236
8.5   Sharing the Task 238
     8.5.1   Architect's Corner 238
     8.5.2   Design Manager's Corner 239
     8.5.3   Unit Designer's Corner 240
8.6   Concluding Remarks 240

References 243

Index 249

Preface

Recent experiences of many chip design companies show that the path to the adoption of formal property verification in the design verification flow is replete with many issues whose answers are still being debated. This includes questions like: *Have I written correct properties? Have I written enough properties? Can I verify every property that I can specify? What should I do if the tool runs into capacity issues? How do I relate what I have verified with my simulation test plan? How should I create an integrated verification plan unifying simulation and formal property verification?*

It is possibly premature to expect definite solutions to all of these questions. This book is an early attempt to address some of these questions and to propose a roadmap for the future.

The focus of this book is not on the core model checking technology for formal property verification. Rather, the book attempts to demonstrate the need of new formal methods that must necessarily supplement the model checking tool in an industrial formal verification flow. These include formal methods for debugging specifications, formal methods for computing coverage, formal methods for scaling the power of property verification tools beyond the limits reached by model checking tools, and formal methods for directing simulation to reach difficult corner case scenarios. This book surveys some of the recent work on these topics and also presents recent research from our group.

Who is this book for? I have attempted to address two aspects in this book – *awareness* of the main issues in adopting property verification technology, and the *formal methods* that help the verification engineer in developing an effective formal verification plan. The contents of this book should be of significant interest to practitioners of formal property verification – design verification engineers responsible for complex chip and system designs: CPUs, DSPs, network processors, graphic processors, and the SOCs that use them. The book also offers new opportunities to CAD researchers, EDA companies and university research groups. I have presented an intuitive introduction to the core formal property verification technology, with the hope that the book may be used in a graduate course on formal property verification.

There is a general suspicion in the chip design community that books on formal property verification are typically terse and the notation used is daunting. In order to reach out to readers without any formal background whatsoever, I have attempted to present most of the concepts in an intuitive (informal) way. I sincerely hope that this attempt has not diluted the theoretical foundations of the subject.

The Formal-V Research Group

The inception of the Formal-V Research Group within the Department of Computer Science and Engineering, Indian Institute of Technology Kharagpur took place in 1997, the same year that Ed Clarke and Moshe Vardi visited our institute. Today this group is a family of 8 Ph.D students, 4-5 graduate students, 2-3 faculty members and several undergraduate students.

We have been fortunate to have several formidable partners in collaborative research. This includes Intel, National Semiconductors, Synopsys, Sun Microsystems, General Motors, and Interra Systems. More information on our research and publications can be found in our home:
http://www.facweb.iitkgp.ernet.in/~pallab/forverif.html

Acknowledgements

The idea of writing this book came from my colleague (previously my Ph.D co-advisor) Prof. P.P Chakrabarti. We set up the Formal-V research group together, and much of our contributions in this subject is due to his vision – something that I have always tried to emulate, but have never succeeded to reach his standards. This book will be incomplete without acknowledging his contribution and help in every aspect – research, development, management, logistics and encouragement.

I have several people to thank for helping us on the research that is covered in this book, and several people to thank in helping me to write this book. Firstly I would like to thank Bijoy Chatterjee of National Semiconductors, Santa Clara for encouraging us to work closely with the industry on problems of industrial relevance. His vision on creating a win-win situation through industry-academia interaction is amazing and most inspiring.

I acknowledge the support of Debashis Chowdhury and Pradip Datta of Synopsys since the early days of the Formal-V group. They have been our true friends in collaborative research and development. I thank Ed Cerny of Synopsys for his comments on some sections of the book.

The most exciting research problems taken up by our group came from Intel. Our research on formal verification coverage (Chapter 5) and on design intent coverage (Chapter 6) was jointly executed with Intel. I sincerely thank Chunduri Rama Mohan of Intel, Folsom for giving us these wonderful research problems, and for his support and participation in the research. It has been a wonderful experience to work with him. I also thank Limor Fix and Roy Armoni for their participation in this research.

This book would not be possible without the contributions of my students. Much of the new research presented in this book is joint work with my students Sayantan Das, Prasenjit Basu, Ansuman Banerjee, Bhaskar Pal and Suchismita Roy. It has been a great privilege to work with such students. Special thanks to Ansuman Banerjee for reviewing the whole of the book and to Bhaskar Pal for reviewing some of the chapters.

I thank Mark de Jongh of Springer for his kind words of encouragement from the beginning of this book project. Working with him on the production of this book has been a truly wonderful experience.

I am very grateful to Erich Marschner of Cadence for reviewing significant portions of the book and for sharing his vast experience with me. His remarks and suggestions have significantly enriched many portions of this book. Interacting with him has been a truly rewarding experience.

I acknowledge the support of my colleagues at the Dept. of Computer Sc. & Engg., I.I.T. Kharagpur for providing the right environment for writing this book.

This book has been developed using LaTeX. Thanks to Don Knuth, Leslie Lamport and many others for gifting this wonderful software to the academic community.

This book is dedicated to my wife, my daughter and my parents – the development of this book significantly encroached upon the time that I should have spent with them. I am grateful to my wife, Sagarika, for her amazing patience, understanding and support – this book would never have been completed without her contribution.

I thank my sister, Dolon and brother-in-law, Anjan for the one month vacation spent at their home in the beautiful city of Dubai during the development of this book. I also thank my in-laws for their help in times of need.

All the illogical statements in this book may be attributed to my twenty month old daughter, Parnika, whose invasions and the ensuing infantile babble with her father may have crept in between the lines of this book.

Finally, whatever little I have been able to achieve is largely due to the everlasting support from my parents.

Kharagpur, India,   *Pallab Dasgupta*
April, 2006.   Associate Professor
                                              Dept. of Computer Sc. & Engg.
                                  Indian Institute of Technology, Kharagpur

1

Introduction

What is formal property verification? A natural language such as English allows us to interpret the term *formal property verification* in two ways, namely:

- *Verification of formal properties*, or

- *Formal methods for property verification*

This inherent ambiguity in natural languages has been the source of many logical bugs in chip designs. Design specifications are sometimes interpreted in different ways by different designers with the result that the design's architectural intent is not implemented correctly. In an era where bugs are more costly than transistors, the industry is beginning to realize the value of using formal specifications.

In practice there are indeed two ways in which property verification is done today. These are static *Assertion-based Verification (ABV)* and dynamic *Assertion-based Verification (ABV)*. In both forms, formal properties specify the correctness requirements of the design, and the goal is to check whether a given implementation satisfies the properties. Static ABV techniques *formally* verify whether *all* possible behaviors of the design satisfy the given properties. Dynamic ABV is a simulation-based approach, where the properties are checked over a simulation run – the verification is thereby confined to only those behaviors that are encountered during the simulation. In this book, we shall refer to static ABV as *formal property verification* (FPV), and continue to use dynamic ABV to refer to the simulation based property verification approach.

The main tasks for a practitioner of property verification are as follows:

1. *Development of the formal property specification.* The main challenge here is to express key features of the design intent in terms of formal properties.

2. *Verifying the consistency and completeness of the specification.* This is a necessary step, because the first task is a non-trivial one and subject to errors and oversights.

3. *Verifying the implementation against the formal property specification.* In order to perform this task effectively, a verification engineer must be aware of the limitations of the verification tool and must know the best way to use the tool under various types of constraints.

All the above tasks are replete with open issues – the focus of this book is to consider some of these issues and attempt to forecast the roadmap for FPV and dynamic ABV within the existing design verification flows of chip design companies. This chapter will summarize some of the major challenges. Let us use the following case as a running example for our discussion.

Example 1.1. Let us consider the specification of a 2-way priority arbiter having the following interface:

mem-arbiter(input r_1, r_2, *clk*, output g_1, g_2)

r_1 and r_2 are the request lines, g_1 and g_2 are the corresponding grant lines, and *clk* is the clock on which the arbiter samples its inputs and performs the arbitration. We assume that the arbitration decision for the inputs at one cycle is reflected by the status of the grant lines in the next cycle. Let us suppose that the specification of the arbiter contains the following requirements:

1. Request line r_1 has higher priority than request line r_2. Whenever r_1 goes high, the grant line g_1 must be asserted for the next two cycles.

2. When none of the request lines are high, the arbiter parks the grant on g_2 in the next cycle.

3. The grant lines, g_1 and g_2, are mutually exclusive.

It is difficult to locate a book on formal verification that does not have an arbiter example - we hereby follow the tradition! □

1.1 Writing Our First Formal Specification

The first task in all forms of property verification is the development of the formal specification. This is a non-trivial task and typically done by a few specialized verification engineers today.

In recent times there have been various efforts within Accellera and the IEEE to define standard property specification languages. These include the definition of the Open Verification Library (OVL), a set of predefined Verilog and VHDL checker modules; the definition of the Property Specification Language (PSL), a language that adds properties and assertions to Verilog, VHDL, SystemVerilog, SystemC, and GDL; and SystemVerilog Assertions (SVA), a subset of the SystemVerilog language that provides a native property and assertion specification capability within SystemVerilog. Although these languages use different syntax, they are closely aligned with respect to semantics. In particular, PSL and SVA were brought into close alignment during their development in Accellera, and that alignment remains between IEEE Standard 1850 PSL and the SVA portion of IEEE Standard 1800 SystemVerilog.

What are in these languages that were not there in earlier languages for specifying constraints? The main feature of these languages, which makes them useful in practice for design verification, is the ability to describe sequences of events over time. The property specification languages derive their formalism from few specific types of logics called *temporal logics*. Temporal logics are extensions of propositional logic, where in addition to the familiar Boolean operators (AND, OR, and NOT) we have *temporal operators* that allow us to specify constraints on the truth of the propositions over time. In other words, temporal logics allow us to specify properties that describe the behavior of a circuit over time, across cycle boundaries.

To get a first-cut workable understanding of temporal logics, let us consider some of the basic temporal operators and their meaning. The operators are interpreted over a state machine – the purpose of verification is to interpret the properties over the state machine representation of the design implementation. The details on temporal operators and property specification languages are given in Chapter 2. We use illustrative examples here to demonstrate their use. The basic set of temporal operators in *Linear Temporal Logic* (LTL) are:

X: *The next-time operator* The property, $X\varphi$, is true at a state of the underlying state machine if φ is true in the next cycle, where φ may be another temporal property or a Boolean property over the state bits. $X\varphi$ is sometimes read as *"next φ"*, and the operator X is called the *next-time operator*.

F: *The future operator* The property, $F\varphi$, is true at a state if φ is true *sometime* (at some state) in the future.

G: *The global operator* The property, $G\varphi$, is true at a state if φ is true *always* in the future.

U: *The until operator* The property, $\varphi \, U \, \psi$ is true at a state if ψ is true at some future state, t, and φ is true at all states leading up to t.

The operators X and U are the only fundamental temporal operators – F and G can be derived from combinations of U and Boolean operators.

Example 1.2. Let us attempt to express the properties of Example 1.1 in LTL. Let us recall the properties:

1. Request line r_1 has higher priority than request line r_2. Whenever r_1 goes high, the grant line g_1 must be asserted for the next two cycles.

2. When none of the request lines are high, the arbiter parks the grant on g_2 in the next cycle.

3. The grant lines, g_1 and g_2, are mutually exclusive.

Typically, the value of a Boolean signal will be treated as TRUE when it is high (or 1), and as FALSE when it is low (or 0). Also we will assume that at most one state transition occurs in each cycle.

The first property may be written as:

$$G[\, r_1 \Rightarrow \, Xg_1 \, \wedge \, XXg_1 \,]$$

The subformula $r_1 \Rightarrow \, Xg_1 \, \wedge \, XXg_1$ says: *If r_1 is high in a state, then g_1 must be true in the next cycle and g_1 must be true in the next next cycle, that is, the second cycle after the initial cycle.* The G operator says that the above subformula must hold on all cycles. This does not mean that r_1 has to be true in all states – those states where r_1 is low satisfy the implication *vacuously*, since $r_1 \Rightarrow \, Xg_1 \, \wedge \, XXg_1$ evaluates to true regardless of the value of g_1 in the next two cycles.

The second property can be written as:

$$G[\, \neg r_1 \, \wedge \, \neg r_2 \, \Rightarrow \, Xg_2 \,]$$

The meaning of this property is exactly as before – *in every state, if both r_1 and r_2 are low, then g_2 must be high in the next cycle.*

The third property can be written as:

$$G[\, \neg g_1 \, \vee \, \neg g_2 \,]$$

This property says: *always at least one among g_1 and g_2 must be low,* which expresses the mutual exclusion requirement. □

Having written our first formal specification, we are now in a position to look at some of the fundamental questions that cross the minds of a practitioner of FPV. The answers to these questions form the major contents of this book.

1.2 Is My Specification Correct?

The first major challenge faced by every verification engineer who uses FPV is to ascertain whether the specification itself is correct. Functional correctness is very difficult to check since we do not have any formal reference against which we may perform this verification. However it is possible to check whether the specification is inherently consistent, or whether there are contradictions within the specification itself. This is a task of considerable importance, but EDA support is not yet adequate.

Example 1.3. Let us examine our first specification. *Is it consistent?* Let us examine the properties again.

$$G[\, r_1 \Rightarrow X g_1 \,\wedge\, XX g_1 \,]$$
$$G[\, \neg r_1 \,\wedge\, \neg r_2 \,\Rightarrow\, X g_2 \,]$$
$$G[\, \neg g_1 \,\vee\, \neg g_2 \,]$$

Let us consider the scenario, where r_1 is high at time t and low at time $t+1$, and r_2 is low at both time steps. The first property requires g_1 to be high at time $t+2$, whereas the second property requires g_2 to be high at time $t+2$ because both r_1 and r_2 are low at $t+1$. The third property prevents both g_1 and g_2 to be asserted at time $t+2$, leading us to a contradiction. Hence we have an inconsistency in our first specification! $\square$

Detecting inconsistencies in specifications is a non-trivial task. It can be modeled as a game between the module and the environment, where the module attempts to satisfy the specification by setting appropriate values to its outputs while the environment attempts to refute the specification by setting values to the input signals. This game alternates between the module and its environment over time – the specification is inconsistent if the environment ever wins. Chapter 4 presents recent research on consistency checking methods for formal property specifications.

Before we proceed further, let us remove the inconsistency from our first specification. The intent of the second property was to specify that g_2 is the default grant. Another way to specify the same intent is:

$$G[\, \neg g_1 \,\Rightarrow\, g_2 \,]$$

This says that g_2 gets the grant whenever we are not committed to give the grant to g_1. Henceforth we will use this property in place of the second property in our specification.

Let us see how this eliminates the problem. If r_1 is high at time t and low at time $t+1$, then the first property requires g_1 to be true at time $t+2$. If g_1 is true, then the above property is vacuously satisfied without requiring g_2 to be true, and hence there is no conflict.

Inconsistencies are very common in design specifications. The realization of the magnitude of this problem will rise as more chip design companies begin to use property verification in their live designs, thereby involving more verification engineers in the task of developing formal specifications. Writing the specifications in formal languages and performing consistency checks can locate some of the most complex inconsistencies in the specifications very early in the design flow, thereby saving a significant quantum of verification (and possible re-designing) effort.

1.3 Have I written Enough Properties?

The popular selling point for FPV is that it is *exhaustive* in nature. Since this guarantee requires the FPV tool to check all possible behaviors of the implementation, it is often misinterpreted as 100% coverage of the design intent. In reality FPV only guarantees that the specified propertiess are verified over all possible behaviors of the implementation – it does not guarantee that the speficied properties are sufficient to cover the design intent. In other words FPV does not verify any part of the design intent that has not been expressed in terms of properties.

One of the main challenges of a verification engineer is to make sure that the formal property specification covers all the correctness requirements of the design. To check the extent of this coverage, we need to compare the set of formal properties with a reference which formally expresses the complete set of correctness requirements. This is an instance of the chicken-and-egg problem, since the reference itself could be used as the formal specification.

It is for this reason that FPV coverage metrics are typically *structural* in nature. In other words, since the property suite represents the first formal functional specification of the design, we do not have any functional reference to compare it with, and therefore we resort to structural coverage. The objective of these structural metrics is to expose gross gaps in the specification. Typically a low value of these metrics indicate that more properties need to be added, but a high value does not necessarily mean that we have high functional coverage. This follows from the fact that low structural coverage almost always means low functional coverage, but the reverse is not always the case.

Most of the existing FPV coverage metrics use a given implementation as the reference for coverage analysis. Chapter 5 outlines some of these metrics. We also present a new style of FPV coverage analysis which is based on the following criticism of existing FPV coverage metrics.

1. If the reference is an implementation, then the coverage of a property suite will change with the implementation. An incomplete implementation may give a false sense of high coverage.

2. We believe that the future of FPV lies in a verification flow, where the properties will be written with the specification document at a time when implementation is yet to begin. At that stage we need to analyze the completeness of the specification in the absence of any reference implementation.

3. Existing coverage metrics have the same evaluation complexity as model checking. Therefore coverage analysis runs into the same capacity issues as model checking.

Our style of coverage analysis compares the property suite against a fault model. Intuitively, if there is any input or non-input signal of the design-under-test which is a *don't care* with respect to every property in the suite, then there is a coverage gap since we did not specify any behavior involving that signal. The details of this approach are presented in Chapter 5.

Example 1.4. Let us again consider our first specification after the modifications in the last section.

$$G[\, r_1 \Rightarrow X g_1 \, \wedge \, X X g_1 \,]$$
$$G[\, \neg g_1 \, \Rightarrow \, g_2 \,]$$
$$G[\, \neg g_1 \, \vee \, \neg g_2 \,]$$

Does this specification cover any behavior where g_1 is required to be high? The answer is yes, the witness being the first property. *But does it ever enforce g_2 to be high?* The answer is no!

This has a serious implication. Consider an implementation that never asserts g_2, and always asserts g_1 regardless of the inputs. None of our properties will be refuted by this implementation and we will be led to believe that the implementation is correct. On the other hand, our coverage analysis will point out that we need to add properties which specify those behaviors where g_2 is forced to be high.

Let us add the following property into our specification:

$$G[\, \neg r_1 \, \wedge \, X \neg r_1 \, \Rightarrow \, X X \neg g_1 \,]$$

Adding this property eliminates the problem. It guarantees that g_1 is never asserted except in those cases covered by the first property. The second property forces g_2 to be high by default.

Let us now look at the input signals. *Do we need to read r_1 at all?* The answer is yes. Suppose g_1 is low at time t. If r_1 is high, the arbiter must assert g_1 at $t + 1$ (by the first property). On the other hand, if r_1 is low, then the aribiter must not assert g_1 at $t + 1$ (by the new property). Therefore it cannot satisfy the specification without reading r_1.

Can we satisfy the specification without reading r_2? The answer is yes! The specification is free from r_2. This is another form of gap which points out the necessity to add properties that cover those cases where the value of r_2 must be considered for setting the correct outputs. Without complicating matters any further, let us accept the fact that in our arbiter specification, r_2 is indeed redundant! □

We are now ready with a consistent and (hopefully) complete specification of our arbiter. The next step is to use this specification to verify the design implementation.

1.4 Property Verification

Having written our first formal specification, we are now faced with the option of using one of the two broad methodologies for property verification, namely *dynamic Assertion Based Verification* (ABV) and *Formal Property Verification* (FPV). Suppose the designer has developed the implementation shown in Fig 1.1. The Verilog code for the module is also given.

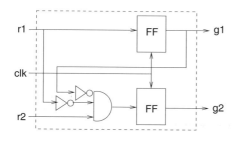

```
module arbiter( r1, r2, g1, g2, clk )
input clk, r1, r2;
output g1, g2;
reg g1, g2;

always @( posedge clk )
begin
    g2 <= r2 & ~r1 & ~g1;
    g1 <= r1;
end
endmodule
```

Fig. 1.1. Arbiter implementation

1.4.1 The Dynamic ABV Approach

In the dynamic ABV approach, we must first write a test bench to drive
inputs into our implementation. The complexity of this task grows rapidly
with the complexity of the design. This is because the environment of a module
is typically constrained by the behavior of the other modules in the design
and by the protocol used for their communication. For example, to verify an
endpoint device in a PCI Express architecture, the test bench must model the
rest of the architecture consisting of other endpoints, the switches and the
root complex. Even after this is done, it is not practically feasible to write
directed tests to sensitize all possible behaviors of this model.

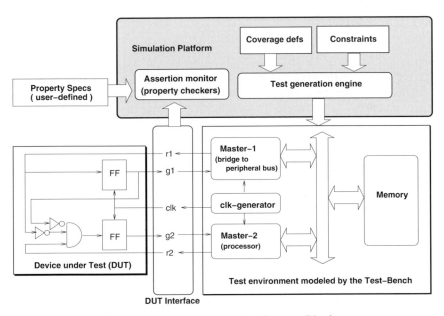

Fig. 1.2. Assertion-based Verification Platform

The dynamic ABV setup is shown in Fig 1.2. We simulate the implementation with the test bench. The assertion checker reads the signals in the
interface and monitors the status of the properties. If any of the properties fail
during the simulation, the checker reports it immediately. The failure points
help the verification engineer to isolate the source of the bug.

There are two key features of dynamic ABV which explain the remarkable
growth in the penetration of this approach. Firstly, it is built over the traditional simulation framework and requires nominal additional effort from the
verification engineer. Secondly, it does not have any major capacity concerns,
since the verification is done over the simulation run.

The main criticism of the dynamic ABV approach is that only those behaviors that are covered by simulation are examined for property violation. The following example demonstrates this problem on our arbiter.

```
module top;
reg clk, r1, r2;
wire g1, g2;

arbiter A(.r1(r1), .r2(r2), .g1(g1), .g2(g2), .clk(clk));

// Clock generator
always #1 clk = ~ clk;

// Model for Master-1
always @(posedge clk)
begin
    r1 = 1;
    @(posedge g1) r1 = 0;
    @(posedge clk); @(posedge clk);
end

// Model for Master-2
always @(posedge clk)
begin
    r2 = 1;
    @(posedge g2) r2 = 0;
end
endmodule
```

Fig. 1.3. Simple Test-Bench fragment for our arbiter

Example 1.5. The implementation of our arbiter is shown in Fig 1.1. The boxes represent flip-flops. The gates have their usual meanings. The verification problem is to determine whether this implementation satisfies our formal specification.

A detailed test environment for our arbiter may consist of the entire memory arbitration environment where it is to be used (see Fig 1.2), including the processor, memory bus, peripheral bridge to the low performance bus containing the peripherals. A simplified test environment for our arbiter consists of a clock generator, and models for the two requesting devices. Fig 1.3 shows a simple test bench for our arbiter[1].

---

[1] Such a simplified model may not model the exact request patterns of the devices – for example, it does not model the memory access latency.

The implementation of our arbiter has a bug. Consider the case when both requests are low for two consecutive cycles. The specification demands that the grant be parked on g_2 (by default), but in our implementation both grants will be low. Though the specification guards against the error, the bug will escape detection because our test bench never creates the relevant cases. □

The example brings out one of the major challenges in property verification. In view of the volume of directed tests that needs to be written in order to achieve a meaningful level of functional coverage, the industry is moving towards coverage driven randomized test generation. This helps in reaching a high level of coverage in short time, but the difficult corner case behaviors are typically left out. Formal properties target these corner case behaviors, but dynamic ABV is not effective unless we can force the test bench to create the relevant scenarios.

The task of manually deriving the tests that cover the scenarios that are relevant to a formal property is a very complex one, even for the seasoned practitioner of ABV. We therefore require automated formal methods that can analyze a property and generate the tests that trigger those properties. Chapter 7 outlines some recent research on this topic.

1.4.2 The FPV Approach

The FPV setup is shown in Fig 1.4. At the heart of this approach we have a *model checking* tool. A model checking algorithm has two main inputs – a formal property and a finite state machine representing the implementation. The role of the algorithm is to search all possible paths of the state machine for a path which refutes one or more properties. If one exists, then the path trace is reported as the counter-example. Otherwise the model checker asserts that the property holds on the implementation.

Example 1.6. Let us again consider the implementation of our arbiter shown in Fig 1.1. The state machine for the arbiter is shown in Fig 1.5. The transitions are labeled by the inputs that enable the transition. The symbol "x" indicates that the value of the signal is a *don't care*.

The objective of model checking is to search for a path in this state machine that refutes one or more properties from our specification. Indeed all paths through the state 00 refute the property:

$$G[\, \neg g_1 \; \Rightarrow \; g_2 \,]$$

The model checker will find a refuting path. Since the path also contains the input sequence for which the refutation occurs, it produces a complete counter-example trace with the appropriate inputs that trigger the incorrect behavior of the module. □

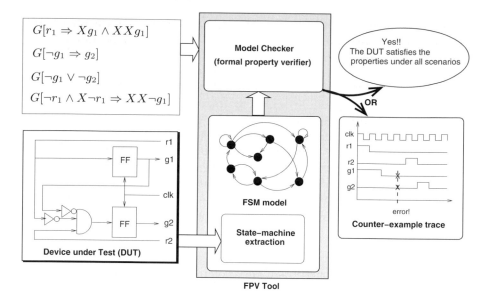

Fig. 1.4. Formal Property Verification Platform

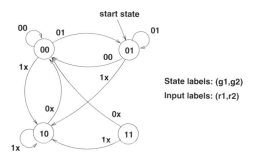

Fig. 1.5. Arbiter State Machine

We outline some of the popular model checking approaches in Chapter 2. The interested reader may refer to [38] for details on model checking methods.

The main limitation of model checking technology is in capacity. There is a popular misconception in this context – verification engineers not conversant with the model checking algorithms often tend to believe that the core model checking algorithms suffer from capacity bottlenecks. This is not exactly the case. Typically the main bottleneck is in the size of the FSM extracted from the implementation. For example, the complexities of both CTL and LTL model checking are linear in the size of this FSM. CTL model checking is also linear in the size of the property. LTL model checking algorithms are exponential in the size of the properties, but there is not much of a capacity issue here, since typical properties written by verification engineers are small in size. The number of states in the machine typically grows exponentially with the number of concurrent components. For designs of moderate size this leads to state explosion.

Most of the recent research in formal property verification has been on developing engineering solutions to the state explosion problem. Some of the notable approaches are:

1. *BDD-based approaches*. Several model checking tools use *binary decision diagrams* to represent the state transition relation succinctly. Typically the size of BDDs tend to grow alarmingly after about 200 variables – hence BDD based approaches face capacity problems on state machines having more than 100 state bits.

2. *Bounded model checking*. Often the verification engineer expects a property to pass or fail within some definite number of cycles. BMC based tools accept a bound on the expected length of a counter-example (if one exists) and unfold the property and the state machine up to the given bound. The clauses generated by this unfolding are fed into a Boolean SAT solver and the result indicates the truth of the property up to the given bound. Since recent SAT solvers are able to handle millions of clauses, BMC can handle larger designs – but the limitation is that we need to know the bound on the length of counter-examples. The bound can be iteratively increased, but the complexity of the SAT instances grow with the increase in the bound.

3. *Abstractions and Approximate Model Checking*. Often the *cone of influence* of a property covers only a fraction of the design. Some of the components of the design may not affect the truth of a property. In such cases it is possible to extract the relevant components and check the property on the reduced state machine. In the other cases also we may choose the components that directly relate to the property and attempt to prove that the property holds in the reduced machine under the assumption that the inputs received from the other components can be arbitrary. If the property holds under this assumption, then obviously it holds on the given design. The reverse is not true, since the inputs under which the property fails may not be asserted by the other components. Hence such methods are referred to as *approximate model checking* methods.

4. *Assume-guarantee Verification*. Existing model checking tools can only handle RTL blocks of small size. These are typically open systems, that is, their behavior is a function of the inputs that they receive from the other components of the design. In order to verify the block in the context of the design, we need to model the behavior of the neighboring components as well, so that the verification considers only meaningful inputs into the block. Assume-guarantee verification is a paradigm that attempts to

discover and model the appropriate assumptions on the input behavior of a block, and to verify the functionality of the block under these assumptions. We attempt to beat the state explosion problem by verifying each block separately – each block is verified under appropriate assumptions on the remaining blocks.

In spite of the advances in the engineering of the model checking tools, it is unlikely that such tools will scale beyond a point because of the magnitude of the underlying state explosion problem. We need a new approach towards formal property verification. We also need a verification flow that demonstrates synergy between the traditional simulation-based verification flow and the formal verification flow.

1.5 Verification by Specification Refinement

As formal property verification tools approach the inherent complexity barrier of the model checking problem, we must investigate other formal methods that can take property verification technology to new heights. We believe that the clue lies in investigating the design flow itself. Formal verification tools are available only recently – designers have been taping out chips for decades without using formal methods. Given the success of the silicon industry and the designer's ability to cope up with Moore's law in the context of growing logical complexity of designs, the number of design errors that have escaped verification have been remarkably few.

One of the cornerstones of this success is the modularity in the design flow. This relates largely to the design architect's ability to hierarchically decompose the functionality of large and complex modules into the functionality of smaller and less complex submodules. At the higher levels of the design flow (RTL or higher), the decomposition leads to design refinement, as implementation specific details are added. The process is continued in the design flow, until the modules are simple enough to be treated as unit level modules.

Today, the decomposition of the architectural specification of a design into the functional specification of the component modules is done manually. The task is non-trivial, and no tool exists that can formally verify the correctness of the decomposition. As a result some of the most complex logical bugs reside in the gap between the architectural specification and the RTL implementation. We believe that this is one of the most important problems before the design verification community.

Recent advances in property specification languages make it possible for the architects to write formal specifications to express the design's architectural intent[2]. Unfortunately, such properties cannot be verified formally on the design due to capacity limitations. Today verification engineers use FPV at the unit level or over small modules – this helps in validating the module against its own specs, but does not solve the larger problem of verifying whether the modules taken together satisfy the global architectural properties.

We believe that the key to solving the problem lies in using a *specification refinement* approach . At the highest level, we start with the formal architectural properties and the design specification (in English). In the top-down design flow, whenever we decompose the design functionality into that of its component modules, we also decompose the formal specification into properties over the component modules. We recursively follow this approach until we reach a level where the modules are within the capacity limits of existing FPV tools.

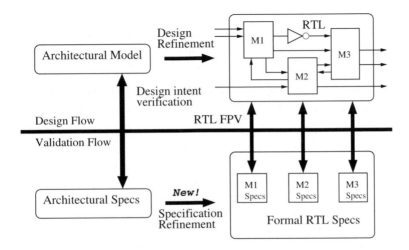

Fig. 1.6. Specification Refinement

The notion of *specification refinement* is not new to current practitioners of FPV. Faced with capacity issues, verification engineers often attempt to check a design property by verifying smaller properties on component modules. Today they lack a formal proof that the component properties together *cover* the original design property, thereby allowing bugs to hide in the gap between the two specifications. In Chapter 6 we present formal methods for verifying this coverage. We either reach a *specification refinement proof*, or we point out the gaps between the two.

---

[2] Today architects are notoriously resistant towards writing formal specifications, but hopefully the picture will change when the benefits of the new approach become apparent.

The *specification refinement checker* is the heart of the specification refinement approach, because it helps us to formally verify the correctness of a specification refinement step, that is, whether *all behaviors that refute the original specification also refute the refined specification.* For example, let $\mathcal{A}$ denote the set of architectural properties of a design. Suppose we implement the design in terms of a set of RTL modules, $M_1, \ldots, M_k$ (Fig 1.6 shows a case with $k = 3$). Since capacity limitations of FPV tools prevent us from verifying $\mathcal{A}$ on the whole design, we manually refine the specification to create sets of properties, $\mathcal{R}_1, \ldots, \mathcal{R}_k$ for the component modules $M_1, \ldots, M_k$. Verifying each M_i with respect to $\mathcal{R}_i$ does not guarantee the conformance of the whole design with $\mathcal{A}$ unless we prove that $\mathcal{R}_1, \ldots, \mathcal{R}_k$ together cover $\mathcal{A}$. This is what a specification refinement proof does for us – it verifies the correctness of the specification refinement, and points out gaps, if any. Since the checker compares specifications (and not a specification versus an implementation), this approach does not have major capacity limitations.

Not all properties are amenable to specification refinement easily, but it is hard to find a case where the design functionality can be decomposed into modules, but the specification cannot be refined.

The specification refinement approach has several associated benefits. Two of the most significant advantages are:

1. *Reuse.* Since verification accounts for more than 70% of the design cycle time, design reuse has marginal benefits unless we can reuse the verification effort. If we simply integrate a pre-designed module with other modules in a new design, we will not save any system level simulation cycles. We believe that in future, every pre-designed module will be accompanied by formal properties that the module is known to guarantee. If we use these properties along with the properties of the other modules to formally prove that the architectural properties are guaranteed, then we will save the additional effort of writing directed simulation tests for the complex behaviors covered by the architectural properties.

2. *Coverage.* Existing RTL FPV coverage metrics are all structural in nature. In the absence of a functional specification that can act as the reference for coverage analysis, most existing FPV coverage metrics use the RTL implementation as the reference. In the specification refinement approach, the higher level functional specification acts as the reference for the lower level specification, and we are able to define a truly functional coverage metric. We discuss this in detail in Chapter 6.

1.6 The New Flow

The end objective of this book is to discover a roadmap for integrating these new and evolving ideas with the existing methodologies for design verification into a unified verification flow. Chapter 8 presents our preliminary ideas on this, based on our interactions with designers, verification engineers, design architects, and EDA companies.

The evolution of a verification flow has its own inertia, and is fraught with hurdles that are not necessarily technical in nature. Our audacity in proposing a new flow is therefore subject to criticism. However, our main aim in proposing the roadmap is to show how the diverse techniques presented in this book can be glued together to suppliment each other, to find the best position of these evolving methodologies in the verification flow, and to answer some of the questions that are being heard regularly from chip design companies who are exploring FPV technology.

Languages for Temporal Properties

Formal verification makes sense only when we have a *formal specification* that acts as the reference for verifying the correctness of a given design implementation. The notion of formal specifications is not new. Fundamentally we are aware that the functionality of all digital circuits may be formally expressed in terms of Boolean functions. For example, a half adder which receives two 1-bit inputs, a and b and produces two 1-bit outputs, namely the sum, s, and the carry c may be specified completely by the Boolean functions:

$$s = (a \ \wedge \ \neg b) \ \vee \ (\neg a \ \wedge \ b)$$
$$c = a \wedge b$$

Given an implementation of a half adder (say, in Verilog RTL) and the above equations, we can formally verify whether the RTL is correct. Typically this is done by translating both the RTL and the Boolean functions into some canonical representation of Boolean functions and then by checking whether the representations are isomorphic. There is a wide range of choices for Boolean function representation, including Binary Decision Diagrams (BDD) [21], Binary Momemt Diagrams (BMD) [22], and ZBDDs [82]. There is also an arsenal of tools and libraries that support these representations, which facilitates the development of formal equivalence checking tools.

Given that the functionality of all digital circuits can be represented by Boolean functions, why do we need these new languages such as PSL and SVA? The reason is fundamental and is very significant towards understanding the basic tenets of formal property verification.

Let us consider the design of an arbiter having request lines r_1 and r_2, and grant lines g_1 and g_2. Suppose we specify the following properties to describe the functionality of the arbiter:

1. Whenever r_1 is raised, the arbiter must assert g_1 within the next two cycles.

2. Whenever r_2 is raised, the arbiter must eventually assert g_2.

3. The grant lines g_1 and g_2 are never asserted together.

Let us now compare the nature of this specification with the one for our half adder. In both cases it is possible to have more than one implementation, which satisfies the specification. Let us consider the following two implementations of the arbiter:

Implementation-1:   The arbiter simply asserts g_1 and g_2 in alternate cycles – regardless of the status of the request lines.

Implementation-2:   Whenever r_1 is raised, the arbiter asserts g_1 in the next cycle. In all other cycles, it asserts g_2.

Fig 2.1 shows the two implementations. It is obvious that these two implementations are not *logically equivalent*, that is, the Boolean functions representing their functionality are not identical. On the other hand, by specifying the Boolean functions for the sum and carry bits of the half adder, we have enforced that every implementation for the half adder must have the same Boolean functionality.

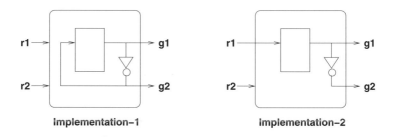

Fig. 2.1. Two implementations of the arbiter

At a high level of abstraction, the design intent is typically expressed in terms of several high-level correctness requirements. Specification of the exact Boolean functionality of the implementation may neither be practical, nor desirable at the high-level. Therefore properties allow us to express a more relaxed version of the specification, covering the critical correctness requirements of the design, but leaving room for design optimization by not specifying the exact Boolean functionality. Recent experience shows that specifying formal properties at the higher levels of the design flow of large and complex chips is both feasible and beneficial – it helps in capturing the essential elements of the design intent in an accurate and non-ambiguous way.

We have not yet justified the need for new languages for formal property specification. One may argue that partial specifications of the Boolean functionality is also a form of formal property specification. This is indeed true. For example, the mutual exclusion between the grant lines of the arbiter can be expressed by the Boolean function: $\neg g_1 \vee \neg g_2$.

On the other hand, let us consider the first property of the arbiter, namely that, *whenever r_1 is raised, the arbiter must assert g_1 within the next two cycles*. This is a property that spans across cycle boundaries. In order to express this property we need the notion of time. The signals r_1, r_2, g_1, g_2, assume different values at different instants of time – the change of values of a signal over time cannot be expressed in terms of the single Boolean variable representing that signal. We may express the property by indexing the signals with a time variable, t, as follows:

$$\forall t \,[\, r_1(t) \;\Rightarrow\; g_1(t+1) \;\vee\; g_1(t+2) \,]$$

$r_1(t)$ and $g_1(t+1)$ denote the value of r_1 and g_1 at time t and $t+1$ respectively. Each time instant, t, describes a state of the signals, r_1, r_2, g_1, g_2, which constitutes the *world* at time t (see Fig 2.2). Intuitively, a property is *temporal* if it involves signals from more than one world.

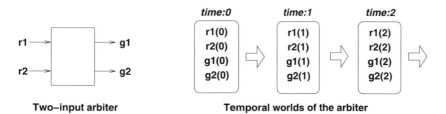

Fig. 2.2. The notion of temporal worlds

The time variable, t, is not a Boolean. Hence the above property is not Boolean. In formal terms, it is not in propositional logic since it contains the first-order variable, t.

We can get rid of the time variable, t, by using two *temporal operators*, namely *next* and *always*. In Chapter 1, we introduced the meaning of these operators. For example, the above arbiter property can be rewritten with these operators as:

$$always \;\; (r_1 \;\Rightarrow\; (next\ g_1) \;\vee\; (next\ next\ g_1)\,)$$

This is a property which contains only Boolean variables, but is not a Boolean function, since it has the new temporal operators. Since the property contains only Boolean variables (propositions), it is a *propositional temporal property*.

The use of temporal logics in verification was proposed by Pnueli in a seminal paper [89]. Since then several different logics have been proposed for specifying temporal properties. All these logics use the two basic temporal operators – *next* and *until*. Some of these logics also use additional temporal operators that can be derived out of the basic two. The logics differ in terms of how we are allowed to mix these operators to express the desired property.

This chapter has two main parts. The first part will introduce the popular temporal operators and the logics that are built around them. In this part we will also introduce some formalisms in an intuitive way that show us how these logics are interpreted over time. In the second part we will look at SystemVerilog Assertions as a language standard and show how it captures the basic temporal operators. We will also study some of its salient features which makes it suitable for specifying design properties.

2.1 The Basic Temporal Operators

The formal introduction to a language has two main parts, namely the *syntax* and the *semantics*. The syntax defines the *grammar* of the language – it tells us how we may construct properties using the basic set of signals and operators. The semantics define the *meaning* of the properties.

The semantics of the traditional temporal logics were defined over *closed systems*, which are finite state machines without any inputs. This tradition has been followed in languages such as SVA and PSL as well – there is no distinction between input and non-input variables in these languages. At this point we will present the semantics of these languages in the traditional form over a non-deterministic finite state machine. Open systems (modules having input bits) can be modeled by treating the input bits also as state bits. This will typically yield a non-deterministic state machine, since the choice of inputs in the next state lies with the environment, and is not a function of the present state.

Suppose J is a finite state machine having k state bits. Each of the 2^k valuations of these state bits represent a *state* of the machine. Let S denote the set of these states. Let R denote the state transition relation of J. R consists of pairs of states, (s_i, s_j), where it is possible to transit from state s_i to state s_j. Finally, J has a start state s. Formaly we say that J is a tuple $\langle S, s, R \rangle$.

Example 2.1. Fig 2.3 shows a 3-bit finite state machine. Let the state bits be n_0, n_1, n_2. The state bits are shown on the nodes. The start state is s_0. Fig 2.3 shows 5 states – the remaining three states are not reachable from the start state and are not shown. The circuit has three outputs, which are functions of the state bits. These are:

$$p = n_0 \ \lor \ n_1$$

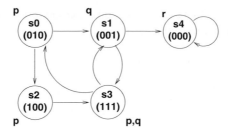

Fig. 2.3. A sample finite state machine

$$q = n_2$$
$$r = \neg n_0 \wedge \neg n_1 \wedge \neg n_2$$

The nodes of Fig 2.3 are labeled by the outputs that are true at that state. We shall use this toy example to demonstrate the meaning of various temporal properties. □

2.1.1 Intuitive Explanation

To convey the semantics of the basic temporal operators, we will first introduce the notion of a *run* (alternatively, a *path* or a *trace*). A run, π, of J is a sequence of states, $\nu_0, \nu_1, \ldots$, where $s = \nu_0$ is the start of the run, and for each i, ν_i represents a state in S, and R contains a transition from the state represented by ν_i to the state represented by ν_{i+1}. In other words, the run is a sequence of states repesenting a valid sequence of state transitions of J. For example the run, $\pi = s_0, s_1, s_3, s_1, s_4, \ldots$, is one run of the state machine shown in Fig 2.3. States of the machine may be revisited in the run – for example we have $\nu_1 = \nu_3 = s_1$ in π. The run, $\pi' = s_0, s_2, s_0, \ldots$, does not belong to this state machine, since it has no transition from s_2 to s_0.

Let us now consider the two fundamental temporal operators, namely *next* and *until*, and a run $\pi = s_0, s_2, s_3, \ldots$.

Next operator: A property, *next f*, is true at a state of a run iff the property f is true at the next state on the run. For example, *next q* is false at the state, s_0, of the run, $\pi = s_0, s_2, s_3, \ldots$, since q is false at the next state s_2. The property *next next q* is true at s_0, of π, because q is true at s_3.

Until operator: A property, *f until g*, is true at a state of a run iff the property g holds on some future state, z, of the run, and the property f holds on all states preceding z on the run. For example, the property, *p until q*, is true at the start state of π, since q is true at the state s_3 and p is true at the states s_0, s_2 preceding s_3 in π. The property, *p until r*, is false on all paths of Fig 2.3, because no r-labelled state can be reached along a p-labelled path starting from s_0.

We will now define the two other operators shown in Chapter 1, namely, *always* and *eventually*. To do this, we need the definitions of the propositions, *TRUE* and *FALSE*. We say that the proposition, *TRUE*, holds in all states, and the proposition, *FALSE*, is false in all states.

Eventually operator: A property, *eventually f*, is true at a state of a run iff the property f holds on some future state in the run. Since the proposition, TRUE, holds on all states, we can express the *eventually* operator using the *until* operator as:

$$eventually\ f\ =\ TRUE\ until\ f$$

The property, *eventually q*, holds on all runs starting from s_0 in Fig 2.3. The property, *eventually r*, does not hold in the run which loops forever in the loop s_0, s_2, s_3, s_0.

Always operator: A property, *always f*, is true on a run iff the property f holds on all states of the run. This is the same as saying that $\neg f$ never holds on the run. In other words we may write:

$$always\ f = \neg eventually\ \neg f$$
$$eventually\ f = \neg always\ \neg f$$

The first equation allows us to express the *always* operator using the *eventually* operator, and in turn, in terms of the *until* operator. The second, says: *sometimes is not never* – there is a seminal paper with this title by Leslie Lamport [78].

The property, *always p* is true in the run which loops forever in the loop, s_0, s_2, s_3, s_0, in Fig 2.3. The property is false in all other runs of the same state machine.

The duality between the *always* and *eventually* operators is not surprising. In fact, it is a variant of DeMorgan's Laws when we interpret the properties over time. This is because:

$$eventually\ f = f\ \vee\ next\ f\ \vee\ next\ next\ f\ \vee\ next\ next\ next\ f\ \ldots$$
$$= \neg\,(\neg f\ \wedge\ next\ \neg f\ \wedge\ next\ next\ \neg f\ \wedge\ next\ next\ next\ \neg f\ \ldots)$$
$$= \neg\,(always\ \neg f)$$

2.1.2 Formal semantics

It is very important to know the formal semantics of a formal property specification language. If the semantics is specified ambiguously, there may be a gap between the property that the designer intends to express and the formal property tool's interpretation of the property that she writes. Bugs may hide in this gap thereby defeating the whole purpose of *formal* property verification. Language lawyer volunteers who make up the working groups of the language standards committees spend years debating over the exact formal semantics of the languages that they standardize. The goal of standardization is to ensure that languages with precise definitions are made available to improve communcation within the industry.

The problem with understanding formal semantics is that they are replete with terse notations. It is widely suspected that the intimidating nature of the notations used in existing literature on formal property verification is one of the main deterants to its wider adoption in practice.

In reality, the formal semantics of the temporal operators do not convey anything more than what we have discussed already. Nevertheless it is worthwhile to present the formal semantics, not only for the sake of completeness, but also to familiarize ourselves with the notations that appear in almost all texts on formal property verification, including some chapters of this book.

To start with, we will use a set of short-forms. We will use X to denote the *next* operator, U to denote the *until* operator, G to denote the *always* operator, and F to denote the *eventually* operator. G means *globally with respect to time*, and F means *in the future*.

Let $\pi = \nu_0, \nu_1, \ldots$ denote a run, and $\pi^k = \nu_k, \nu_{k+1}, \ldots$ denote the part of π starting from ν_k. We will use the notation $\pi \models f$ to denote that the property f holds on the run π. Given a run π, we will also use the notation $\nu_k \models f$ to denote $\pi^k \models f$. In other words, a property is said to be true at an intermediate state of the run iff the fragment of the run starting from that state satisfies the property. The formal semantics of the basic temporal operators are as follows:

- $\pi \models Xf$ iff $\pi_1 \models f$

- $\pi \models f \ U \ g$ iff $\exists j$ such that $\pi_j \models g$ and $\forall i, 0 \le i < j$ we have $\pi_i \models f$.

Fg is a short-form for TRUE U g, and Gf is a short-form for $\neg F \neg f$.

2.2 Logics for Temporal Specification

Temporal logics tell us how we can create complex temporal properties by putting together one or more temporal operators. There are broadly two classes of these logics, namely *linear time logics* and *branching time logics*. Linear time logics allow the specification of properties over linear traces or runs of a finite state machine – intuitively, we say that the property holds on the machine if it holds on all runs of the machine. Branching time logics allow the specification of properties over the computation tree created by a state traversal of the state machine.

2.2.1 Linear Temporal Logic

Designers and validation engineers typically express and interpret the RTL in terms of the simulation semantics of the HDL. They are accustomed to verifying the correctness of the RTL by checking certain behaviors over simulation traces. Therefore it is not suprising that linear time logics have been the natural choice for design validation, and form the backbone of most existing property specification languages, including Forspec, PSL and SVA.

Linear Temporal Logic (LTL) is the most popular among linear time logics. We can define the language recursively as follows:

- All Boolean formulas over the state variables are LTL properties.

- If f and g are LTL properties, then so are: $\neg f$, Xf, and $f \; U \; g$.

We can also use the short-forms, Fg for $true \; U \; g$, and Gf for $\neg(true \; U \; \neg f)$.

The semantics of LTL is as follows. We will say that the property f holds on a state machine, J, iff f holds on all paths of the state machine starting from its start state. The semantics of f on a path is as defined in the last section.

Let us see some sample LTL properties obtained by using one or more temporal operators. We refer to Fig 2.3.

- The property $p \; U \; q$ is true in the state machine, since all paths from s_0 satisfy this property.

- The property Fq is true in the state machine, but the property GFq is not true. This is because we have the path, $\pi = s_0, s_1, s_4, \ldots$, which does not satisfy Fq from s_4 onwards.

- The property $p \; U \; (q \; U \; r)$ is not true in the state machine, because it may get trapped in the loop, s_0, s_2, s_3, s_0.

Fig 2.4 shows some sample LTL properties and some sample runs that satisfy these properties.

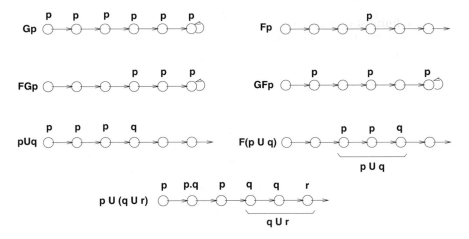

Fig. 2.4. Some sample LTL properties

2.2.2 Computation Tree Logic

Computation Tree Logic (CTL) is a branching time temporal logic. Properties described in this logic are interpreted over the *computation tree* obtained by unfolding the state machine as a tree. We will elaborate on this shortly, but let us first study the basic features of this logic.

CTL has two *path quatifiers* in addition to the usual temporal operators. These are the *existential* path quantifier E, and the *universal* path quantifier A.

- The property $A\varphi$ is true at a state, ν, iff φ is true on *all* runs starting from ν.

- The property $E\varphi$ is true at a state, ν, iff φ is true on *some* run starting from ν.

In CTL we have the restriction that every subformula of the form Xf, Gf, Fg, and $f\,U\,g$ must be prefixed by an E or A. Therefore, we may define the language as:

- All Boolean formulas over the state variables are CTL properties.

- If f and g are CTL properties, then so are: $\neg f$, AXf, EXf, $A[f\,U\,g]$ and $E[f\,U\,g]$.

We also have the usual short-forms Fg for $true\,U\,g$, and Gf for $\neg(true\,U\,\neg f)$. Consequently, EFg, AFg, EGf, AGf are CTL properties. Fig 2.5 shows some

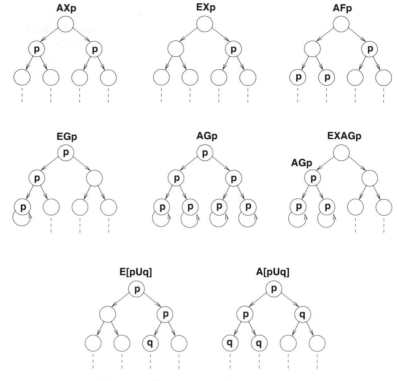

Fig. 2.5. Some sample CTL properties

sample CTL properties and some sample computation trees that satisfy these properties.

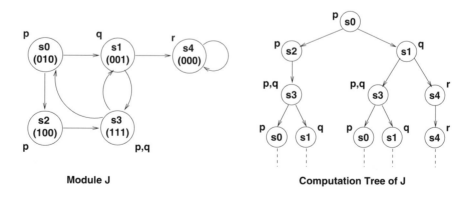

Fig. 2.6. The notion of a Computation Tree

What is the significance of a *computation tree*? Let us consider the state machine of Fig 2.3. We can unfold the state machine into the infinite tree shown in Fig 2.6. Each path of this tree is a run or a *computation* of the state machine. CTL allows us to define properties over the nodes of this computation tree.

For example, consider the property:

$$A[p \ U \ E[q \ U \ r]]$$

This property is true at the start state s_0 of our state machine. This follows from the fact that $E[q \ U \ r]$ is true at s_1 and s_3 (since there is a q-labeled sequence of states to the r-labelled state $\dot{s}_4$), and every path from s_0 reaches one of these states through p-labelled states. It is not possible to express this property in LTL.

2.2.3 LTL versus CTL

Can all LTL properties be expressed in CTL using the universal path quantifier, A? The answer is, no. For example, the LTL property, FGp, is not equivalent to the CTL property, $AFAGp$. Fig 2.7 shows an example which satisfies FGp but does not satisfy $AFAGp$.

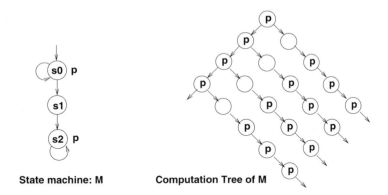

State machine: M   **Computation Tree of M**

Fig. 2.7. Module M satisfies FGp, but not AFAGp

When do we use a linear time logic, and when do we use a branching time logic? This is a matter of considerable debate, and is hardly agreed upon. However, experience shows that linear time logics are the natural choice for black-box testing. For example, while specifying the behavior of a module, we can write linear time properties over its interface signals without knowing the internal state machine of the module.

On the other hand, branching time logics are useful for verifying properties over a given state machine. For example, when we develop an automotive control system we may typically model the control system as an abstract state machine, verify whether this control system satisfies certain safety properties and then expand the abstract state machine into the actual control system. When a branching time property fails, we must interpret the failure in terms of the actual states of the system, which is not possible in black-box testing.

The temporal logic CTL* combines the expressive power of linear and branching time temporal logics. CTL and LTL are fragments of CTL*. There has been several interesting extensions of these languages that demonstrate the tradeoff between the expressive power of the language and the complexity of model checking (that is, formally verifying) properties specified in these languages.

2.2.4 Real Time Temporal Logics

The temporal operators discussed so far, namely *next* (X), *until* (U), *always* (G), and *eventually* (F), are *temporal* because they can define sequences of events over time. Significantly, none of these operators with the exception of the *next* operator, attempt to *quantify* time. For example the property, *eventually* f, requires f to be true in future, but does not specify any time bound by which f needs to be true.

Real time temporal operators are intuitively simple extensions of the basic *untimed* temporal operators where we annotate the operator with a time bound. The real time extensions of CTL and LTL simply use these bounded operators (as well as the unbounded ones).

The bounded Until operator: The property $fU_{[a,b]}g$ is true on a run, $\pi = s_0, s_1, \ldots$, iff there exists a k, $a \le k \le b$ such that g is true at s_k on π, and f is true on all preceding states, $s_0, \ldots, s_{k-1}$. Formally,

$$\pi \models f\, U_{[a,b]}\, g \text{ iff } \exists k, a \le k \le b, \nu_k \models g \text{ and } \forall i, 0 \le i < k \text{ we have } \nu_i \models f$$

The bounded LTL property $p\, U_{[1,3]}\, q$ is true at the state s_0 of Fig 2.3. The bounded CTL property:

$$A[p\, U_{[1,3]}\, E[q\, U_{[1,2]}\, r]]$$

is also true at s_0. This is because s_3 and s_1 satisfy $E[q\, U_{[1,2]}\, r]$ (since they can reach s_4 within the time bound $[1, 2]$), and we will reach s_1 or s_3 along all paths from s_0 within the time bound $[1, 3]$.

The bounded Eventually operator: The property $F_{[a,b]}g$ is true on a run, $\pi = s_0, s_1, \ldots$, iff there exists a k, $a \le k \le b$ such that g is true at s_k on π.

For example, the bounded LTL property $F_{[1,3]}q$ is true at the state s_0 of Fig 2.3. The bounded CTL property $EF_{[2,4]}r$ is true at s_0. The property $F_{[a,b]}g$ is equivalent to the bounded-until property, $true\ U_{[a,b]}\ g$.

The bounded Always operator: The property $G_{[a,b]}f$ is true on a run, $\pi = s_0, s_1, \ldots$, iff f is true in every state in the sequence, $s_a, \ldots, s_b$. The bounded CTL property $EG_{[3,9]}q$ is true in the state s_0 of Fig 2.3 – consider the run from s_0 through s_2 which alternates between s_3 and s_1 at least 8 times. The bounded LTL property $G_{[0,1]}\neg r$ is true at s_0 since no run can reach s_4 is less than 2 cycles.

Real time operators are extremely useful in practice. Most design properties have a well defined time bound, and must be satisfied within that time.

Since the real time operators deal with finite bounds, a and b, they can be expressed in terms of the X operator. For example, the property $F_{[2,4]}q$ can be rewritten as:

$$F_{[2,4]}\ q\ =\ XX(q\ \vee\ Xq\ \vee\ XXq)$$

and $p\ U_{[3,4]}\ q$ can be rewritten as:

$$p\ U_{[3,4]}\ q\ =\ (p\ \wedge\ Xp\ \wedge\ XXp)\ \wedge\ XXX(q\ \vee\ (p\ \wedge\ Xq))$$

The first part of the property specifies that p be must be true in the present cycle and the next two cycles. The second part of the property specifies that q must be true in the third cycle, failing which, p must be true in the third cycle and q must be true in the fourth cycle.

Therefore, real time operators actually help us to succinctly express properties that would require too many X operators otherwise.

2.3 SystemVerilog Assertions

The success of property verification in the industry depends to a large extent on the evolution of language standards for property specification. The task of building language standards to be followed by many companies is one of the hardest tasks, and not entirely for technical reasons. Currently *SystemVerilog Assertions* (SVA) [102], *Property Specification Language* (PSL) [92], and *Open Verification Library* (OVL) [85] are the three main alternatives for property specification. SVA is the natural choice for designers using SystemVerilog; PSL is a good choice if one is working with VHDL, Verilog or SystemC; and OVL is a good choice if one is not willing to learn either SVA or PSL.

OVL, PSL and SVA were all developed initially under Accellera [3]. At this point, PSL has become an IEEE standard (IEEE 1850 PSL), and SVA is part of the IEEE 1800 SystemVerilog standard.

The syntax and semantics of SVA and PSL have minor differences. Therefore, it is equivalent to study any one of the two for formal property specification. We choose SVA in this book for the following reasons:

1. SVA is part of the SystemVerilog language – which is a design and verification language. We will demonstrate how property specification is glued with the design and test bench language to create an integrated verification platform. This platform will be a key component of our proposed formal verification roadmap.

2. Some of our tools outlined in this book are based on SystemVerilog. For several reasons our research group has been inclined towards SystemVerilog as a design and verification language.

An excellent overview of PSL is given in the book, *Applied Formal Verification*, by Perry and Foster [88].

This section is not intended to be a comprehensive overview of the SystemVerilog Assertions language. The complete syntax and semantics is given in the SystemVerilog LRM. Our goal in this section is to expose the reader to some of the interesting modeling features of the language, and to share some of our experiences in developing verification IPs with this language. Our presentation is based on SystemVerilog 3.1a.

2.3.1 A Quick Overview

We will start with our 2-way priority arbiter described in Chapter 1. The arbiter has the following interface:

mem-arbiter(input r_1, r_2, clk, non-input g_1, g_2)

r_1 and r_2 are the request lines, g_1 and g_2 are the corresponding grant lines, and clk is the clock on which the arbiter samples its inputs and performs the arbitration. We developed the following LTL properties for the arbiter in Chapter 1:

$$P1: \quad G[\, r_1 \Rightarrow Xg_1 \,\wedge\, XXg_1 \,]$$
$$P2: \quad G[\, \neg g_1 \Rightarrow g_2 \,]$$
$$P3: \quad G[\, \neg g_1 \vee \neg g_2 \,]$$
$$P4: \quad G[\, \neg r_1 \wedge X\neg r_1 \Rightarrow XX\neg g_1 \,]$$

We will start by writing these properties in SVA. Then we will integrate these properties into an assertion-based verification framework. We will use the (incorrect) arbiter implementation (in Verilog) of Chapter 1 for this purpose.

Each temporal property describes a *sequence* of events. In SVA we capture *sequences of events* through *sequence expressions*. For example, the property P1, which says that *whenever r_1 is high, g_1 must be high in the next two cycles*, can be written in SVA as follows:

$$\text{r1} \ |-> \ \text{\#\#1 g1 \#\#1 g1}$$

The symbol, $|->$, represents the implication operator.

The ##1 operator can be used to implement the X (*next*) operator of LTL. ##1 g1 is true at a state if g1 is true in the next state of the run. The sequence:

$$\text{\#\#1 g1 \#\#1 g1}$$

is true at a state if g1 is true in the next two cycles.

Since the semantics of LTL is defined over a state machine, the clock is implicit in LTL properties. In practice, a circuit may have several clocks. Also we may define a property over signal values which are sampled only at the occurrence of a specific event. To provide this flexibility, SVA allows us to define the sampling clock for a property. The property P1 may therefore be written as:

```
property P1;
    @(posedge clk)
    r1 |-> ##1 g1 ##1 g1 ;
endproperty
```

The expression, r1 $|->$ ##1 g1 ##1 g1, is evaluated at every posedge of clk, which was our intent while using the G (*always*) operator in the LTL formula for P1. The property P2 can be similarly written as:

```
property P2;
    @(posedge clk)
    !g1 |-> g2 ;
endproperty
```

The NOT operator, !, denotes the negation operator, $\neg$, of Boolean algebra. We will use a separate operator for negation of SVA properties.

The property P3 can be written in SVA as:

```
property P3;
    @(posedge clk)
    !g1 || !g2 ;
endproperty
```

The OR operator, ||, is used over Boolean expressions. We will use a separate operator for logical OR between SVA properties.

The property P4 shows an interesting difference between the semantics of SVA and LTL. We may write P4 in SVA as:

```
property P4;
    @(posedge clk)
    !r1 ##1 !r1 |- > ##1 !g1 ;
endproperty
```

We have the sequence, !r1 ##1 !r1, in the antecedent of the implication. This sequence matches at time $t + 1$ iff r1 is false at t and $t + 1$. The property specifies that g1 must be false at time $t + 2$ whenever the sequence matches at time $t + 1$. Therefore the matching of the consequent of the implication begins *after* the matching of the antecedent succeeds. Compare this with the LTL property:

$$G[\ \neg r_1 \ \wedge \ X\neg r_1 \ \Rightarrow \ XX\neg g_1\ ]$$

If r_1 is false at time t and $t + 1$, the antecedent of the implication evaluates to true. The consequent $XX\neg g_1$ is defined with respect to time t, from which we *started to match* the antecedent, that is, we expect g_1 to be false at $t + 2$ (specified by the two X operators). On the other hand in the SVA property, the consequent, ##1 !g1, is defined with respect to time $t + 1$, that is, the time at which we successfully *completed the matching* of the antecedent. This explains the use of a single ##1 operator in the consequent part of the SVA property, as opposed to the use of two X operators in the consequent part of the LTL property.

Having written our properties in SVA, we must now *bind* the properties with the RTL for the arbiter module. SystemVerilog provides the notion of an *interface* to facilitate this task. The interface defines the set of signals of the module that are *visible* to the test-bench or the assertion checker. For example, in our arbiter the interface consists of the signals, r_1, r_2, g_1, g_2. We define the properties in the interface and then use them as *assertions*.

The definition of the interface and assertions for the arbiter is shown in Fig 2.8. Each assertion has a name, an assert statement, and a clause which indicates the action to be taken when the assertion fails. For example, consider the assertion:

```
Mutex:
    assert property(P3)
    else $display("Property P3 has failed");
```

The name of the assertion is mutex. This name is typically used by the assertion checker to refer to the assertion (say, when it fails or passes vac-

```
interface ArbChecker(    input g1,
                         input g2,
                         input r1,
                         input r2,
                         input clk ) ;

property P1;
    @(posedge clk) r1 |− > ##1 g1 ##1 g1 ;
endproperty

property P2;
    @(posedge clk) !g1 |− > g2 ;
endproperty

property P3;
    @(posedge clk) !g1 || !g2 ;
endproperty

property P4;
    @(posedge clk) !r1 ##1 !r1 |− > ##1 !g1 ;
endproperty

GrantWhenRequest:
    assert property(P1)
    else $display("Property P1 has failed");

OneGrantHigh:
    assert property(P2)
    else $display("Property P2 has failed");

Mutex:
    assert property(P3)
    else $display("Property P3 has failed");

NoGrantWhenNoRequest:
    assert property(P4)
    else $display("Property P4 has failed");

endinterface
```

Fig. 2.8. SVA interface and property definitions for the arbiter

uously). The statements following the `else` clause are executed when the assertion fails (typically these are non-synthesizable statements). In this case we use this clause to print an error message.

```
module Top;

wire r1;
wire r2;
wire g1;
wire g2;
reg clk;

// Instantiation of the module arbiter
arbiter A(r1, r2, g1, g2, clk);

// Clock generator
initial begin
    clk = 1;
    forever begin
        #1 clk = ~ clk;
    end
end

// Rest of the test bench code ...

endmodule
```

Fig. 2.9. Structure of the test bench

Fig 2.9 shows the structure of the test bench for the arbiter. We need to *bind* the interface, `ArbChecker`, with the arbiter. One way to do this is to *bind* the interface with the test bench using the following statement.

```
bind Top ArbChecker ArbC( g1, g2, r1, r2, clk )
```

Binding is an important step in assertion-based verification. It associates the names of the signals used in our properties with the names of the corresponding signals in the RTL module. This ability to create an association between the propositional variables in the properties and the RTL variables used in the module enables us to delineate the task of creating a verification IP for a design from the task of writing the RTL. Today, numerous verification IPs for standard protocols, such as PCI, PCI-XP, AMBA, USB, etc are available off-the-shelf. The validation engineer only needs to bind these property suites with the corresponding signal names in the design implementation in order to check these properties. Fig 2.10 shows the complete picture.

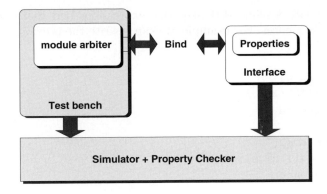

Fig. 2.10. Binding a verification IP

2.3.2 The Notion of Sequences

Sequence expressions are the basic building blocks of SVA. Some typical examples of sequence expressions are as follows.

```
##0 r1          // r1 is true in this cycle
##1 r1          // r1 is true in the next cycle
##5 r1          // r1 is true exactly after 5 cycles
##[5:9] r1      // r1 is true sometime between the 5th and 9th cycle.
```

It is easy to see the equivalence between these and the bounded LTL operators of Section 2.2.4. For example:

```
##1 r1          is the same as        Xr1
##5 r1          is the same as        F[5,5]r1
##[5:9] r1      is the same as        F[5,9]r1
```

The sequence expression, `r2 ##3 g2`, matches if $r2$ is high and after 3 cycles $g2$ is high. In other words, `r2 ##3 g2`, is equivalent to the bounded LTL property, $r2 \wedge F_{[3,3]}g2$.

Sequence expressions can be used to express sequences of signal values. For example, consider the sequence:

$$a \ \#\#[1:5] \ (b||c) \ \#\#3 \ d$$

The sequence matches when we have a followed by b or c after k cycles followed by d after 3 cycles, where k is any integer between 1 and 5. This property is equivalent to the bounded LTL property:

$$a \ \wedge \ F_{[1,5]}[\ (b \ \vee \ c) \ \wedge \ F_{[3,3]}d \ ]$$

We can associate a name with sequence expressions and then use them to define other sequences. For example, we can rewrite the previous property as follows.

```
sequence s1;
    (b||c) ##3 d;
endsequence

sequence s2;
    a ##[1:5] s1;
endsequence
```

Sequence $s2$ uses sequence $s1$. The ability to define sequences in this way is very useful in practice. Complex temporal events can be defined using sequences, and then these can be used in properties that are triggered by these events.

2.3.3 Sequence Operations: Repetition

Sequences of events over time may often be expressed by regular expressions. For example, the LTL property, $p\ U\ q$, may be expressed by the regular expression, p^*q, which matches every run where we have zero or more p-satisfying states followed by a q-satisfying state.

SVA supports the notion of regular expressions through three repetition operators. These are:

1. *Consecutive repetition.* The sequence, p [*5], matches when five consecutive states satisfy p. Therefore, the sequence p [*5] ##1 q, matches when 5 consecutive states satisfy p followed by a state satisfying q. This is similar to the bounded LTL property, $p\ U_{[5,5]}\ q$.

   The sequence, p [*3:5] ##1 q, matches when k consecutive matches of p is followed by a match of q, where k is an integer between 3 and 5. This is similar to the bounded LTL property, $p\ U_{[3,5]}\ q$.

   To specify an unbounded number of repetitions, we may use the dollar sign ($). The sequence, p [*3:$] ##1 q, matches when k consecutive matches of p is followed by a match of q, where k is any finite integer greater than or equal to 3.

   Suppose we wish to express the property: *whenever the request r2 of our arbiter is raised, the arbiter must eventually assert g2, and the requesting device must hold the request line at high until the grant arrives.* We can express this intent as:

$$\texttt{r2 } |->\texttt{ r2 [*1:\$] ##1 g2}$$

The LTL property, $p \ U \ q$, is equivalent to: p [*0:\$] ##1 q.

2. *Goto repetition.* Consecutive repetition enables us to count consecutive occurrences of an event. Sometimes we want to count events which may not occur in consecutive cycles. For example, suppose a slave splits a burst transfer when it becomes non-available during the ongoing transfer, and we wish to abort a transaction if more than two splits occur during the transaction. The property may be expressed in SVA as:

$$\texttt{split [*->2] ##1 abort}$$

The two split events may not be consecutive. The abort event occurs one cycle after the second split event.

The sequence, p [*->5] ##1 q, matches at time t, if q matches at time t, and p matches 5 times before time t, *including one at time $t - 1$*. The last phrase is important, and is the main difference with *non-consecutive repetitions*.

The sequence, p [*->3:5] ##1 q, matches at time t, if q matches at time t, and p matches between 3 to 5 times before time t, including one at time $t - 1$.

3. *Non-consecutive repetition.* Non-consecutive repetitions are similar to Goto repetitions with a minor difference. For example, our expression:

$$\texttt{split [*->2] ##1 abort}$$

required the *abort* event to happen exactly one cycle after the second *split* event. On the other hand, the same expression using the non-consecutive repetition operator:

$$\texttt{split [*=2] ##1 abort}$$

specifies that the abort event must occur *sometime* after the second split event – not necessarily in the next cycle.

The sequence, p [*=3:5] ##1 q, matches at time t, if q matches at time t, and p matches between 3 to 5 times before time t.

The repetition operators can be applied on more complex sequence expressions as well. For example, the expression:

$$\texttt{(r1 ##1 g1) [*3] } |->\texttt{ ##1 g2}$$

specifies that *whenever three consecutive alternations of r1 and g1 occur, it is followed by a grant to the default slave, g2.* This sequence is equivalent to:

```
(r1 ##1 g1 ##1 r1 ##1 g1 ##1 r1 ##1 g1) |- > ##1 g2
```

The repetition operators significantly contribute to the expressive power of SVA. They are also one of the main sources of errors in property specifications. Arbitrary nestings of repetition operators must be avoided for the sake of clarity and readability of the formal specification.

2.3.4 Sequence Operations: AND, INTERSECT, OR

The semantics of logical operations over temporal properties is not obvious from their Boolean counterparts. This is because logical operations on temporal properties are interpreted over runs, and different temporal properties may match at different points of time.

Given an AND / INTERSECT / OR over two sequence expressions, s_1 and s_2, the main issue is to define the time point at which we may return success or failure. For example, if x and y are boolean variables, then the Boolean property, x && y, is true at all states where both x and y are true. On the other hand, consider two sequences, s_1 and s_2, where s_1 is x ##2 y, and s_2 is x [*0:$] z. Then the sequence expression, s_1 and s_2, can match in many ways, such as:

```
(x && z) ##2 y
x ##1 z ##1 y
x ##1 x ##1 (y && z)
x ##1 x ##1 (y && x) ##1 (x [*0:$] z)
```

In other words, the run must satisfy both s_1 and s_2. The end of a successful match is the earliest point of time when both sequences have matched. A failure can happen much earlier. For example, if the first state does not satisfy x, then we have an immediate failure.

Similarly, the sequence expression, s_1 or s_2, holds on all runs that satisfy s_1 or s_2 (or both). The end of a successful match is the earliest point of time when at least one among s_1 and s_2 have matched. A failure can be detected only after both s_1 and s_2 have failed.

The **intersect** operator is a variant of the **and** operator, where both sequence expressions must match starting at the same cycle and ending at the same cycle. Therefore if s_1 is x ##2 y, and s_2 is x [*0:$] z, then the sequence expression, s_1 **intersect** s_2, can match in only one way, namely:

```
x ##1 x ##1 (y && z)
```

2.3.5 Local Variables

Let us consider the FIFO queue shown in Fig 2.11. The queue stores integers. The module raises the *QFull* signal when the queue is full. In order to push a number into the queue when *QFull* is low, we float the number on the *DataIn* lines and raise the *Put* signal. When the queue is full, we flush the queue by performing a sequence of pop operations. In each pop operation, we raise the *Get* signal and read the data from the *DataOut* lines.

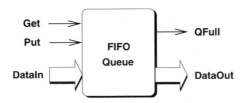

Fig. 2.11. A FIFO Queue

We can express the FIFO property as follows. *If X and Y are any two data items such that X was pushed before Y, then X will come out of the queue before Y.* In this property, the values of X and Y are not important – they are placeholders for any two data items entering and leaving the queue.

The concept of *local variables* in SVA properties enable us to express such properties formally. For example, we can express the FIFO property in SVA as follows.

```
property FIFO_check;
    int x;
    int y;
    @(posedge clk)
    ((Put && !QFull, x = DataIn) ##[1,$]
        (Put && !QFull, y = DataIn)) |->
            ##[1,$] ((Get && x == DataOut) ##[1,$]
                (Get && y == DataOut)) ;
endproperty
```

The variables x and y are local variables for this property. In order to match the property the checker will attempt all possible *instantiations* of these variables on the run. For example, if the queue size is 3 and we insert 2, 5 and 3, before flushing the queue, then the checker will match the antecedent in three ways, namely $(x = 2, y = 3)$, $(x = 2, y = 5)$, and $(x = 5, y = 3)$. If the sequence in which data is popped is 2, 3 and 5, then the checks for the first two cases will match, while the check for the case $(x = 5, y = 3)$ will fail and the report will show this failure case.

2.3.6 Properties

Properties specify a behavior of a system. We can use a property as an `assert`, `assume` or `cover`. We will explain the use of each of these types in the next two sections.

In the arbiter example, we declared the properties inside the interface, ArbChecker. Properties can be defined at other places also (such as inside a module definition), but it is good practice to define the properties inside the interface definition if we wish to reuse the properties. For example, the properties used for verifying whether a master device is compliant with the PCI Bus protocol may be defined within the master interface definition. Then the properties will be reused automatically for all PCI masters that use instances of the same interface.

SVA supports several operators for defining properties from sequence expressions. These include, `not`, `and`, `or`, `if` ... `else`, and the implication operators, $|->$ and $|=>$. The semantics of the operators, `and` and `or`, are similar to those for sequence expressions. The operator, `not`, represents negation over temporal properties – `not` P is true on a run iff P fails on the run. This operator is not the same as the ! operator which is used for negation over Boolean formulas.

Suppose we wish to express the following property in SVA: *If the request line r1 goes high, then in the next cycle, the grant line g1 must go high and r1 must be de-asserted, otherwise in the next cycle, the grant g2 must go high.* We may express this property in SVA as:

```
property P;
    @(posedge clk)
    if (r1) then ##1 (g1 && !r1)
        else ##1 g2 ;
endproperty
```

SVA does not allow the conditional expression of the `if`-statement to be a sequence expression. For example, we may want to express the property that, *whenever the request line r2 is not granted for two consecutive cycles, the request is lowered in the next cycle* as:

```
property ThisIsNotOkay;
    @(posedge clk)
    if (r2 ##1 (!g2 && r2) ##1 !g2) then ##1 !r2 ;
endproperty
```

This is not allowed by SVA, because the sequence expression, r2 ##1 (!g2
&& r2) ##1 !g2 is not supported as a conditional for the if-statement. In
such cases, we may use the implication operator as follows:

```
property ThisIsOkay;
    @(posedge clk)
    r2 ##1 (!g2 && r2) ##1 !g2 |-> ##1 !r2 ;
endproperty
```

The antecedent of the implication operators can be a sequence expression,
but it cannot be a property expression. For example, the following property
is not supported:

```
property WrongAgain;
    @(posedge clk)
    (a |-> ##1 b) |-> ##1 c ;
endproperty
```

This is not supported because the antecedent of the second implication
operator is not a sequence expression.

SVA supports two types of implication operators, namely $|->$ and $|=>$.
The semantics of these operators are almost the same except for the following
difference:

- In the property, s1 |-> s2, the match of $s2$ starts from the same cycle as
  the one in which we complete a match for $s1$.

- In the property, s1 |=> s2, the match of $s2$ starts from the cycle *after*
  the one in which we complete a match for $s1$.

Not surprisingly, the first operator is called the *overlapped* implication operator, while the latter is called the *non-overlapped* implication operator.

One interesting feature of SVA property specifications is the use of the
disable iff clause. Suppose we expect the arbiter to *service the request $r2$*
by asserting $g2$ within the next 32 cycles. However if the requesting device
lowers $r2$ before receiving $g2$, then the arbiter no longer needs to service the
request. We may write this property in SVA as:

```
property UseOfDisableIff;
    @(posedge clk)
    disable iff (!r2) r2 |-> ##[1:32] g2 ;
endproperty
```

The match of the sequence, r2 |− > ##[1:$] g2, starts at every cycle where $r2$ is true. The property can be satisfied in two ways:

- $g2$ is asserted within 32 cycles of $r2$, or

- $r2$ is lowered before 32 cycles and before $g2$ is asserted. This returns a (*vacuous*) match because the property matching is disabled when !$r2$ becomes true.

The property can fail in only one way – $r2$ remains asserted for the 32 cycles, but $g2$ is not asserted.

The common use of the `disable iff` clause is in aborting the matching of a property when an interrupt occurs. For example, consider the following property:

```
property DisableOnReset;
    @(posedge clk)
    disable iff (reset) x |− > ##[1:16] y ;
endproperty
```

The property requires y to be asserted within 16 cycles after the occurrence of x, *except when reset is asserted in between*.

The `disable iff` clause is sometimes used in verification IPs to allow the user to turn selected properties on/off. For example, suppose the designer of a PCI XP based system has not yet implemented flow control in the system. In order to verify the existing system with a verification IP for PCI XP, the validation engineer must turn off all properties related to flow control. If the architect of the verification IP uses a common flag variable in `disable iff` clauses to guard all flow control properties, then the validation engineer can simply set/reset the flag value to turn the flow control properties on/off. Since the actual properties are not visible to the user of the verification IP (because of IP reasons), this mechanism has almost the same effect as commenting out the unwanted properties (with a minor overhead).

2.3.7 Assume-Assert Specifications

Since the semantics of temporal logics are defined over closed systems, there is no syntactic difference between input and output signals. However, the behavior of most modules that we verify in practice is a function of the inputs received from the environment of the module. For example, when we define the compliance requirements for a master device with the PCI Bus protocol, we assume that the other devices in the system (which includes the Bus arbiter, slave devices, and other master devices) work correctly, that is, they drive PCI

Bus compliant inputs into the module under test. In other words, we expect the module to satisfy the PCI Bus properties correctly in a valid PCI Bus based system, not in every arbitrary environment.

A formal specification for an open system such as a module must therefore include the *assumptions* that we make about the environment of the module under test – *assumptions* under which we expect the module to satisfy certain properties.

SVA allows us to specify `assume` properties to describe the assumptions about the environment, and `assert` properties to describe the properties that must be *guaranteed* by the module under the given *assumptions*. This style of reasoning is called *assume-guarantee* reasoning.

As an example, let us return to our priority arbiter example. Suppose we add the property: *every low priority request, r2, is eventually granted by the arbiter by asserting g2*. We can write this property as:

```
property NoStarvation;
    @(posedge clk)
    r2 |- > ##[1:$] g2 ;
endproperty
```

This property will not hold under all circumstances. This is due to the property, P1:

```
property P1;
    @(posedge clk)
    r1 |- > ##1 g1 ##1 g1 ;
endproperty
```

If $r1$ is never low for two consecutive cycles, then P1 requires $g1$ to be asserted forever, which in turn will starve the low priority device. Adding property *NoStarvation* into our specification will make the specification self-conflicting and inconsistent.

Now suppose we are given that whenever $g1$ is asserted, $r1$ remains low for the next 4 cycles. We can express this as another property:

```
property FairnessOfr1;
    @(posedge clk)
    g1 |- > ##1 (!r1) [*4] ;
endproperty
```

The above property cannot be guaranteed by the arbiter, because it has no control over its input, $r1$. However if we expect the request pattern to honor this requirement, then we can treat this property as an *assumption* on the environment behavior. Under this assumption, the property, *NoStarvation*, is consistent with P1 and actually expresses the design intent. SVA allows us to express the desired intent as follows:

```
property FairnessOfr1;
    @(posedge clk)
    g1 |- > ##1 (!r1) [*4] ;
endproperty
AssumeR1IsFair: assume property (FairnessOfr1);

property NoStarvation;
    @(posedge clk)
    r2 |- > ##[1:$] g2 ;
endproperty
AssertNoStarvation: assert property (NoStarvation)
    else $display("Low priority device is starving");
```

There are several notable points here:

1. The `assume` directive is used to specify that the property is an assumption. The `assert` directive is used to specify that the property is an assertion.

2. Both the `assume` and the `assert` properties have input and output variables of the module. A common misconception among early users is that `assume` properties may only use input signals.

3. The `assume` properties are not bound to any specific `assert` property. The `assume` properties express assumptions about the environment behavior regardless of what is expected from the module under test.

4. In dynamic assertion based verification, both the `assume` and the `assert` properties need to be checked during simulation. If the `assume` property fails, then the check for the `assert` property may simply be aborted, since the `assert` property is required to hold only when the `assume` property holds.

5. In formal property verification, the `assume` properties are interpreted as constraints under which the `assert` properties must be checked. In other words, if A is an `assume` property and B, C and D are `assert` properties, then the formal tool will attempt to check the property, $A \Rightarrow B \land C \land D$ on the design implementation. We will discuss the implications of `assume` properties on the capacity issues of FPV tools in Chapter 3.

`assume` properties can also be used to specify the value ranges of inputs as well as the relative probabilities of each value. This feature is very useful for generating random tests. We will pick up this issue again in Chapter 7.

2.3.8 The Cover Statement

Writing assertions for specific scenarios is not sufficient to expose bugs by dynamic property verification unless the simulation comes up with the relevant scenarios that sensitize the bug. It is therefore an important objective to check whether the properties that we have written are actually interpreted non-vacuously during simulation. For example, consider the property:

```
property P4;
    @(posedge clk)
    !r1 ##1 !r1 |− > ##1 !g1 ;
endproperty
```

This property is interpreted non-vacuously only when $r1$ is low in two consecutive cycles, that is, when we have a match for the sequence, !r1 ##1 !r1. Formally, vacuity is defined in SVA only for properties having the implication operators – a property is satisfied non-vacuously only if the consequent part of the property has a role in it. Whenever the antecedent fails, the property matches vacuously.

One of the main benefits of the **cover** directive of SVA is that it gives us the number of non-vacuous interpretations of an assertion during simulation. For example, suppose we specify the **cover** property:

```
property P4;
    @(posedge clk)
    !r1 ##1 !r1 |− > ##1 !g1 ;
endproperty
cover property (P4)
```

The results of the coverage statement for this property will show the number of times the property was attempted, the number of times it succeeded, the number of times it failed, and the number of times it succeeded because of vacuity.

2.4 Architectural Styles for Assertion IPs

The task of deriving a set of formal properties from the English language specification of a design or a protocol is a non-trivial task having many different considerations. As a result, off-the-shelf *assertion IPs* for many standard protocols, such as PCI Bus, AMBA Bus, IBM Coreconnect and PCI XP, are in demand, and are available from multiple vendors. Assertion IPs consist of assertion suites and interface definitions that are used to bind the assertions with the design under test.

An assertion IP is an integral part of a *verification IP*. A verification IP typically consists of an assertion IP and a collection of models that may be used to construct the environment for the design under test. For example, to create a realistic test environment for a PCI XP endpoint device, we need models for the switches, the root complex, and other endpoints. While using a verification IP, the validation engineer will be able to build a test environment by instantiating one or more such models in the test bench. Fig 2.12 shows the complete picture.

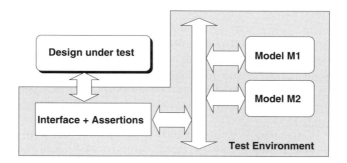

Fig. 2.12. Test environment using a verification IP

More recently several methodologies have been proposed for developing verification suites. These include the e Reuse Methodology, RVM for Open Vera and VM for SystemVerilog. These verification methodologies provide guidelines for structuring the test environment using models, defining the coverage points, and monitoring the progress of coverage driven constrained random test generation.

What are the steps and the issues in developing the core assertion IP? The steps are intuitively simple, but the issues are many. For example, the issues in developing an assertion IP for a formal (static) property verification tool are quite different from that for dynamic property verification. Data path properties such as verifying whether a transfer protocol uses big endian or little endian memory addressing is not a major issue for dynamic property verification, but may run into capacity issues for a FPV tool. Most of the existing styles are evolving, and it is possibly premature to expect a single undisputed development style. Nevertheless, we shall highlight some of the design choices in the main steps. These are purely our own insights based on developing several verification IPs, such as ARM AMBA Bus, IBM Coreconnect, Hypertransport and PCI XP.

2.4.1 The Main Steps

The development of an assertion IP typically starts with a design/protocol specification. The tasks involved in building an assertion IP are as follows – these are not necessarily executed in the given order.

1. *Identifying the functional properties.* The complexity of this task varies largely with the clarity and details provided in the design specs. For example, we found the ARM AMBA Bus specification to be particularly well documented. This was more of an exception than the rule, since design specs often have lots of ambiguity and this accounts for a significant fraction of the logical bugs in the design.

2. *Identifying the interface.* This represents the task of identifying the set of signals that should be made visible to the assertion monitor. In other words, these are the signals over which the assertions will be written. In some cases this task is not obvious.

   For example, let us consider the task of CRC checking on packets in the PCI XP protocol. In order to execute the CRC check we need the whole packet, but the packet is broken down into words and transmitted over many cycles. PCI XP has several properties that define the behavior when a CRC check fails. There are two options for checking these properties:

   - We locate the signal that indicates CRC check match/fail in the destination endpoint, and use that signal to trigger the assertions that relate to actions taken on CRC failure. In this case, we need the single Boolean signal at the interface.

   - We make the *receive*-buffer at the destination endpoint visible at the interface. In this case the interface definition should call the CRC function to determine whether the buffer contents are valid, and then use the result to trigger the appropriate properties.

   In the first case, we are implicitly assuming that the CRC checker at the destination endpoint is functionally correct. In the second case the CRC check is part of the verification effort. The first approach is clearly more amenable for formal verification, since the interface is light-weight – but the functional correctness of the CRC checker at the destination endpoint must be verified separately.

3. *Coding the assertions.* The same set of properties may be expressed in different ways in a given language. We will study some coding styles later in this section.

4. *Evaluating the consistency of the assertion IP.* Since assertion IP development is a non-trivial task, mistakes are common. In Chapter 4 we will

study some of the common forms of inconsistencies, and formal methods for detecting such inconsistencies.

5. *Evaluating the completeness of the assertion IP.* One of the main challenges of formal property verification comes from bugs which like to hide between the design intent and the formal specification. We will present formal methods for evaluating the coverage of an assertion IP in Chapter 5 and Chapter 6.

In the remainder of this Chapter we will focus on the styles for coding assertions in an assertion IP.

2.4.2 Coding Styles: *Event and State-based Checkers*

The easiest way to present the architectural styles for assertion IPs is through an illustrative example. The relative benefits of these styles become apparent when one attempts to code a specification having many properties. Nevertheless we shall make an attempt to convey the basic ideas through a toy example.

The MyBus Protocol

We consider a simple Bus protocol that supports multiple master devices, multiple memory-mapped slave devices, and a single arbiter. The Bus has 64-bit multiplexed address and data lines. During a transfer, address and data are time multiplexed on these 64 lines, with address and data appearing alternately in address and data cycles respectively. We will focus on the behavior of master devices. The master interface is shown in Fig 2.13.

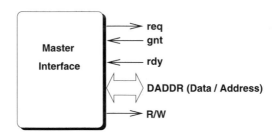

Fig. 2.13. Master interface for the MyBus protocol

A master interface is IDLE when the master device does not intend to perform a transfer. When the master intends to start a transfer, it raises its request line, **req**, and waits for the arbiter to return a grant, **gnt**.

On getting a `gnt`, the master floats the address in the Bus and waits from the next cycle onwards for the `rdy` signal from the slave device. We refer to this phase of the transfer as the ADDRESS cycle.

The `rdy` signal from the slave indicates that the slave is ready for the transfer – on receiving this signal the master enters the DATA cycle and does the following:

1. In the case of a write transfer, it floats the data on the Bus.

2. In the case of a read transfer, it expects the slave to produce the data on the Bus.

The intent to read/write is indicated by a R/W signal – high indicates write intent, low indicates read intent. In both cases, the slave is expected to lower the `rdy` signal in the data cycle and raise it again in the next cycle or whenever it is ready for the next transfer. After each data cycle, the master may start another address cycle by floating the next address on the Bus. At any point of time the master can return to the IDLE state by lowering the `req` line, which signals the end of the transfer to the arbiter. A sample transfer is shown in Fig 2.14.

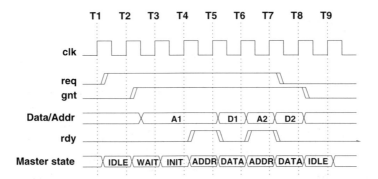

Fig. 2.14. A sample transfer for the MyBus protocol

We choose the following properties of the MyBus protocol for our analysis.

1. The protocol is non-preemptive. Once granted, the master owns the Bus until it lowers its `req` line.

2. If the master is in the ADDRESS cycle, it should not change the address floated in the Bus until it receives the `rdy` signal from the slave.

3. Each DATA cycle is of unit cycle duration.

Let us start by coding these properties directly in SVA. The first property can be expressed as:

```
property NoPreemption;
    @(posedge clk)
    $rose(gnt) |- > ##1 gnt [*1:$] ##0 !req ;
endproperty
```

The predicate `$rose(gnt)` is true in a cycle if the signal `gnt` rose in that cycle.

The triggering condition for the second property is that the master must be in the ADDRESS cycle. *How do we express that the master is in the AD-DRESS cycle?* This is a non-trivial question. For example, we may attempt to express the intent through the following property:

```
property IncorrectAddressStable;
    int x;
    @(posedge clk)
    (req && gnt && !rdy, x = DADDR)
            |- > ##1 (x == DADDR) ;
endproperty
```

The property says that the address lines must remain stable when the master has control of the bus (given that `req` and `gnt` are high), but the slave is not ready (that is, `rdy` is low).

On closer inspection of Fig 2.14 we find that the antecedent of the above property is not quite correct. This is because the protocol multiplexes the address and data lines, and the antecedent matches at T6 also. In order to differentiate between the ADDRESS cycle and the DATA cycle, we may write the property as:

```
property AddressStable;
    int x;
    @(posedge clk)
    (req && gnt && !rdy && !$fell(rdy), x = DADDR)
            |- > ##1 (x == DADDR) ;
endproperty
```

The predicate `!$fell(rdy)` excludes T6 from being treated as an AD-DRESS cycle.

The third property requires us to identify the DATA cycle and express that the master does not spend consecutive cycles in this phase. The master

is in the DATA cycle when **gnt** is high, and **rdy** fell in this cycle. We intend
to express that this combination is never true on two consecutive cycles.

```
property SingleCycleDataTransfer;
    @(posedge clk)
    (gnt && $fell(rdy)) |-> ##1 (!gnt || !$fell(rdy)) ;
endproperty
```

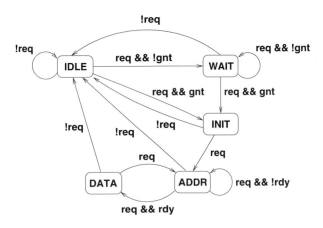

Fig. 2.15. State Machine for the MyBus Master Device

There are several problems in writing properties in this way, particularly
the last two.

1. The antecedent parts of the implications become large and complex. This
   is because the antecedent typically refers to some state of the protocol, and
   expressing that state in terms of the signal values is not straightforward.
   In many cases, the signal values do not uniquely identify the state of the
   protocol, and one has to encode recent history (as sequence expressions)
   in the antecedent in order to uniquely indicate the state at which the
   property becomes applicable.

2. The properties are not readable. Looking at the SVA code for the third
   property, it is hard to read the intent of the property, unless explicit
   documentation proves that the expression, **gnt && $fell(rdy)**, identifies
   a DATA cycle.

3. Since the antecedent parts of the implications are not obvious, errors in
   coding the properties are common and debugging is hard.

In order to get around the above problems of such an *event-based* coding of
properties, validation engineers employ a *state-based* coding of the properties.

In this approach, we first create an abstract state machine for the protocol and code it in SVA. The assertions are then coded in terms of these states.

Fig 2.15 shows an abstract state machine for the MyBus master device. Fig 2.16 shows the structure of the master interface definition with an encoding of this state machine in SystemVerilog.

The state machine contains only sufficient information that carries it through the major phases of the protocol. For example, the status of the **gnt** lines are not indicated in the DATA and ADDR cycles, though this is relevant to the first property.

```
interface MasterInterface(      input req,
                                input gnt,
                                input rdy,
                                input int DADDR,
                                input clk ) ;

    logic [2:0] state;

    'define IDLE 3'b000
    'define WAIT 3'b001
    'define INIT 3'b010
    'define ADDR 3'b011
    'define DATA 3'b100

    always @ (posedge clk)
    begin
        case (state)
            'IDLE: state <= req? (gnt? 'INIT : 'WAIT) : 'IDLE ;
            'WAIT: state <= req? (gnt? 'INIT : 'WAIT) : 'IDLE ;
            'INIT: state <= req? 'ADDR : 'IDLE ;
            'ADDR: state <= req? (rdy? 'DATA : 'ADDR) : 'IDLE ;
            'DATA: state <= req? 'ADDR : 'IDLE ;
        endcase
    end


    ----
    // Property and Assertion definitions ....
    ----

    initial begin
        state = 'IDLE;
    end
endinterface
```

Fig. 2.16. State machine model defined inside an Interface

We can now use the state machine to define our properties. The first property remains as it is. The second and third properties are rewritten as follows:

```
property AddressStable;
    int x;
    @(posedge clk)
    (state == 'ADDR, x = DADDR) |-> ##1 (x == DADDR) ;
endproperty

property SingleCycleDataTransfer;
    @(posedge clk)
    (state == 'DATA) |-> ##1 !(state == 'DATA) ;
endproperty
```

The antecedents of the implication are now intuitively simple, and close to the English language statements of the properties. For large and complex protocols, this turns out to be a significant benefit.

The main challenge in this approach is in creating the state machine model for the protocol – specifically in deciding the level of details that should go into the state machine model. If we make the state machine too detailed, then the coding of the state machine will become complex, unreadable and error prone. If we make it too abstract, we will not be able to simplify the coding of the properties. Therefore a hybrid approach seems to be the best choice.

Typically the state machine model of a protocol can be factorized into a set of simple state machine models. For example, the state machine of Fig 2.15 may be factorized into two state machines, as shown in Fig 2.17. The first state machine describes the state of the Bus access in terms of the req and gnt lines. The second state machine is triggered by the outedge from the INIT state of the first, and describes a more detailed state of the master during the transfer.

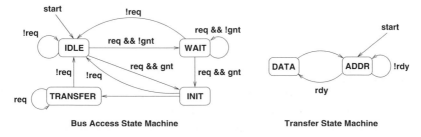

Fig. 2.17. Factored State Machines for the MyBus Master

Suppose we encode the first state machine in terms of the 2-bit variable, *state1*, and the second state machine in terms of the 1-bit variable, *state2*. We may rewrite the properties again as follows:

```
property AddressStable;
    int x;
    @(posedge clk)
    (state1 == 'TRANSFER && state2 == 'ADDR, x = DADDR)
            |- > ##1 (x == DADDR) ;
endproperty

property SingleCycleDataTransfer;
    @(posedge clk)
    (state1 == 'TRANSFER && state2 == 'DATA)
            |- > ##1 !(state2 == 'DATA) ;
endproperty
```

The advantage of the factorized state machine is that the individual state machines are very simple. It allows us to use only the relevant signals in the factored machine.

We found factored state machine models to be very useful when the protocol contains multiple concurrent activities which overlapped in time. For example, in most modern Bus protocols the address and data lines are different (not multiplexed) and the address and data phases are pipelined during a transfer – the address for the next transfer is floated at the same time when the previous data is being read/written. In such cases it helps to model the status of the address lines and the status of the data lines as separate state machines.

2.5 Concluding Remarks

This chapter presented the basics of formal property specification. Languages such as SVA and PSL are based around a core set of temporal operators. The operators are quite simple, but the way in which we allow them to be combined dictates both the expressive power of the language, as well as the complexity of checking properties specified in that language.

A basic exposure to the underlying temporal logics builds the foundation for learning the exisiting property specification languages and their future derivatives. In other words, the difference between static constraints on variables and temporal constraints on the way a variable is allowed to change over time, is a fundamental concept that helps in developing the style of property specification. Like any other new language paradigm, this basis is not built overnight, but cultivated through experience. The remainder of this book will attempt to establish that the expected returns from this effort is rich and promising in terms of reducing overall validation effort.

One question is being pondered upon by many chip design companies – *Who should spend the effort of mastering the art of formal property specification?* The answer possibly lies in finding out who benefits most from property verification. Off course everyone benefits if the chip has fewer errors and faster turn-around time, but this common goal does not typically unify the finer dynamics within the chip design flow. Micro-architects are often notoriously resistant towards writing formal properties, designers do not often consider themselves to be a part of the validation flow, validation engineers are often not sure whether the properties that they code actually mean what they expect them to mean. We will return to these issues in Chapter 8.

Existing property verification tools have to be supported by an arsenal of new formal tools that will enable the validation engineer to evaluate the formal specification – in terms of *correctness* and *completeness*. We will pick up these challenges in Chapter 4 and Chapter 5.

2.6 Bibliographic Notes

The notion of temporal reasoning is not new. Temporal logics were originally developed by philosophers for reasoning about events across temporal worlds using natural language [69]. Pnueli was the first to use temporal logics to reason about the correctness of concurrent progams [89]. This was also the first time that Linear Temporal Logic was used for reasoning about the temporal behavior of programs.

In [35], Clarke, Emerson and Sistla proposed the use of Computation Tree Logic for the specification and verification of branching time properties. This was also one of the first papers to present algorithms for temporal logic model checking.

The notion of extending the basic temporal operators with quantitative constructs was introduced by Emerson, Mok, Sistla and Srinivasan in [53]. In this paper, the authors introduced the logic RTCTL, which is the real time extension of CTL. Extensions to dense real time temporal logics was presented by Alur, Courcoubetis and Dill in [4, 5, 6].

The adoption of temporal logics in design validation started with several independent developments of property specification languages for facilitating the specification of formal design properties. Roy Armoni and others developed a language called Forspec [10] for Intel, and proceeded to build a complete formal verification tool suite around this language. The language PSL originates from the language called Sugar developed at IBM labs at Haifa, Israel [101]. Synopsys developed a language called Open Vera Assertions [84] for formal property specification. The notion of using property libraries for facilitating the task of specifying properties was also born with the OVL libraries [85].

Powered by a consortium of chip design and EDA companies, several technical committees within Accellera have contributed towards the arduous task of developing language standards for property specification languages. The Accellera Formal Verification Technical Committee chaired by Harry Foster and Erich Marschner developed the standards for PSL [92], which is now being adopted as an IEEE standard. Foster and Marschner were also involved in the development of SVA as a part of the Accellera SystemVerilog standards, which again has been adopted by IEEE. OVL [85] is a collection of libraries standardized by the Accellera OVL technical committee – for use by engineers who do not want to write properties in SVA or PSL.

3

How Does the Property Checker Work?

In the early stages of the inception of formal property verification, one question is heard quite often among design validation engineers – *Do I need to know the algorithms for formal property verification in order to use the technology?*

Given the state of existing FPV technology, the answer to this question is not entirely negative. Fortunately, what is required is a basic awareness of how the FPV tool works. Formal verification is not a push-button solution to the validation problem. It is a methodology - and the right methodology enables the validation engineer to get around the many limitations of the FPV tools. For example, some of the frequently asked questions are:

- Why does the tool get into capacity issues? What are my options if this happens on a given design?

- What is a bounded model checker? When do I go for it?

- What are BDD-based and SAT-based tools? Which one do I want?

- How does the tool use the *assume* constraints during FPV?

The objective of this chapter is to provide a general idea about the FPV algorithms so that the answers to the above questions become apparent. The intention is to present the methodology without introducing too many formalisms. Readers interested in more details on the working of these formal methods are referred to the excellent text by Clarke, Grumberg and Peled on model checking [38].

Since our goal is a general awareness of the issues in FPV techniques, we will restrict our attention to the basic temporal operators. This will enable us to study the basic difficulties of model checking without getting into the intricacies of verifying a more complex language such as SVA or PSL.

In this chapter, we will continue with the arbiter example presented in Chapter 1. Let us recall the arbiter example.

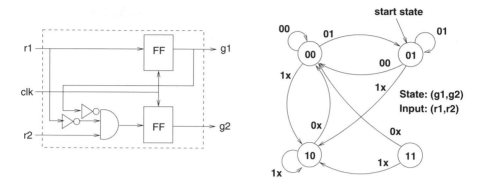

Fig. 3.1. Arbiter implementation and state machine

Example 3.1. We consider the specification of a 2-way priority arbiter having the following interface:

mem-arbiter(input r_1, r_2, *clk*, non-input g_1, g_2)

r_1 and r_2 are the request lines, g_1 and g_2 are the corresponding grant lines, and *clk* is the clock on which the arbiter samples its inputs and performs the arbitration. We developed the following properties for the arbiter in Chapter 1:

> P1: $G[\, r_1 \Rightarrow Xg_1 \,\wedge\, XXg_1 \,]$
> P2: $G[\, \neg g_1 \,\Rightarrow\, g_2 \,]$
> P3: $G[\, \neg g_1 \,\vee\, \neg g_2 \,]$
> P4: $G[\, \neg r_1 \,\wedge\, X\neg r_1 \,\Rightarrow\, XX\neg g_1 \,]$

We also developed an implementation for our arbiter. The implementation and the state machine of the implementation are shown in Fig 3.1.

It is easy to see (informally) that the implementation has at least two bugs:

- It refutes P1 when r_1 is high for only one cycle. The arbiter asserts g_1 for only one cycle, where as P1 requires g_1 to be high for two consecutive cycles.

- If r_1 and r_2 are both low at some time, then P2 fails in the next cycle since both g_1 and g_2 are low.

In this Chapter we will demonstrate the way in which such bugs are detected by the property verification techniques. □

3.1 Checkers are State Machines!

The first important observation in formal property verification is that the property checker needs to maintain its state. In other words, based on what it has seen so far, the checker has a state that determines what needs to be checked now (in the present cycle) and what needs to be checked in the future.

Consider the property P1 in Example 3.1. Whenever r_1 is asserted, the checker needs to verify $Xg_1 \wedge XXg_1$, that is, it must check g_1 in the next two cycles. We can describe the state of the checker at a given point t in terms of the property it needs to check from t onwards. For example, if r_1 is true at time t, then it needs to check $Xg_1 \wedge XXg_1$ at t, it needs to check $g_1 \wedge Xg_1$ at $t+1$, and it needs to check g_1 at $t+2$.

Similarly, consider the property P4 in Example 3.1. If at time t, we have $\neg r_1$ (that is, r_1 is low), then the state of the checker at time t can be described by the property, $X\neg r_1 \Rightarrow XX\neg g_1$. If r_1 goes high at $t+1$, then the property is satisfied. Otherwise, the state of the checker at time $t+1$ can be described by the property, $X\neg g_1$. At time $t+2$, this leads to failure if the arbiter asserts g_1, and success otherwise.

A checker may therefore be represented by a state machine. The first step of formal property verification is to create this state machine from the property. We will outline the methodology for this task through the following property:

$$\varphi : \qquad p\ U\ (q\ U\ r)$$

Formally the input to the automaton is the state of the module under test (that is, the current values of the signals) and the state of the automaton encodes what needs to be checked henceforth. The set of states, S, of the automaton consists of all subformulas of the given property, φ, and their negations.

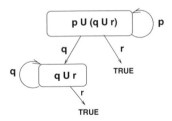

Fig. 3.2. Automaton for the property: $p\ U\ (q\ U\ r)$

For example, the subformulas of φ includes, $p, q, r, q\ U\ r$, and φ itself. A simplified automaton for this property is shown in Fig 3.2. Initially the

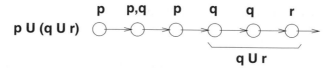

Fig. 3.3. A run for the property: $p \ U \ (q \ U \ r)$

checker needs to check the whole property $p \ U \ (q \ U \ r)$ – hence the initial state is the root state. At this state we have the following possibilities:

1. If the module asserts r, the property is satisfied, and we reach the TRUE state.

2. If the module does not assert r, then we have three cases.

   a) The module asserts q, but not p. In this case, we need to check $q \ U \ r$ from the next state onwards – hence the checker automaton moves to the state labeled by $q \ U \ r$.

   b) The module asserts p, but not q. In this case, we need to check $p \ U \ (q \ U \ r)$ from the next state onwards – hence the checker automaton stays at the root state.

   c) *What happens if the module asserts both p and q?* In this case, can it simply look for $q \ U \ r$? The answer is negative. Consider the valid run shown in Fig 3.3. In the second state of this run, both p and q are true. If we abandon the check for $p \ U \ (q \ U \ r)$ and only check for $q \ U \ r$ from the second state onwards, then we will reach failure (incorrectly) in the third state, since neither r nor q is true at the third state. Therefore, the checker needs to check both $p \ U \ (q \ U \ r)$ as well as $q \ U \ r$ from the second state. This represents a non-deterministic choice between staying in the state labeled by $p \ U \ (q \ U \ r)$ and moving into the state labeled by $q \ U \ r$.

State machines for properties, such as the one shown in Fig 3.2, are formally known as *weak alternating automata* (WAA). Specifically, it has been shown that a special class of WAAs, known as *very weak alternating automata* (VWAA) represent exactly the class of properties that can be defined using LTL formulas[74].

There are standard algorithms for converting a non-deterministic finite automaton into a deterministic finite automaton. The deterministic equivalent of the non-deterministic VWAA for a property is exactly the desired checker automaton.

What is the size of the checker automaton for a property? This is an important question for estimating the capacity of several formal methods used in FPV and related tasks. To answer this question we need to recall two important facts:

1. The states of the automaton consists of all subformulas of the property. Therefore, the number of states of the non-deterministic automaton is linear in the length of the property.

2. The number of states in the deterministic version of a non-deterministic automata can be exponential in the number of states of the non-deterministic automaton. In other words, each state of the deterministic automaton represents a *set of subformulas* of the original property, namely the set of subformulas that needs to be checked from that state onwards.

The above observations lead us to conclude that the size of the deterministic checker automaton is exponential in the length of the property. If our property involves large bit vectors (such as data and address buses), then this figure can become prohibitive.

3.2 The Verification Strategy

The task of verification may be interpreted as a search for a run of the implementation that refutes the property. If we find such a run, then we have found a bug.

The checker automaton for a property accepts all runs that satisfy the property. Therefore we create the checker automaton for the negation of the given property. This automaton accepts all runs that refute the original property. Our objective is to check whether the implementation has any of these runs. This can be done in broadly two ways:

1. Co-simulate the implementation with the checker automaton and report a bug whenever the checker automaton for the negation of the property reaches the TRUE state. This is what we do during dynamic property verification.

2. Take the formal product of the state machine of the implementation with the checker automaton of the negation of the property. If the product is *non-empty*, that is, it contains at least one run, then we have found a bug. Otherwise the implementation is guaranteed to satisfy the property. This is the methdology followed in static or formal property verification.

In the next few sections we will elaborate the above approaches with examples. We will also examine the main engineering issues that determine the feasibility and scalability of these approaches.

3.3 Dynamic Property Verification

Dynamic property verification is making significant penetration in pre-silicon validation of digital chip designs. The advantage of this approach is that it does not suffer from major capacity limitations. Consequently it is being adopted for the verification of system level architectural properties. The flip side of this approach is that it covers only as much of the behaviors as are exercised by the simulation test bench. A given error will not be detected in a given dynamic property verification run if the scenario in which it occurs is not exercised in that run.

Why does dynamic property verification not suffer from serious capacity issues? An intuitive understanding of the property monitor that checks the properties during simulation will lead us to the answer.

Let us consider the following approach for dynamic property verification for a given property φ:

1. Create the checker automaton for $\neg\varphi$ and create its deterministic equivalent.

2. Start the simulation of the implementation with the checker automaton at the start state.

3. After each simulation step, determine the next state of the checker automaton by presenting the signal values at the end of the simulation step.

4. If the checker automaton reaches an accepting state then report the existence of a bug. Otherwise, proceed to the next simulation step.

The main additional overhead in this approach is in the first step, where we create the checker automaton and then create its deterministic version. If the length of the property is large (say, it has large bit-vectors), then the checker automaton can be large.

Actually we can do better. We can create the checker automaton on-the-fly. At the end of each simulation cycle we only require the possible next states of the checker automaton. Since we get to know the signal values at the end of each cycle, we can use the information to prune one or more next states of the checker automaton.

How can we create a checker automaton on-the-fly? The answer lies in the following simpler question: *If we are to check a property φ at a given time t, what do we need to check at time $t + 1$ for a given valuation of the signals at time t?* The answer to this question recursively gives us an answer to the former question.

Temporal properties can be recursively re-written in terms of Boolean propositions that must be satisfied by the current values of the signals and temporal properties that must hold in the next state. The rewriting rules are as follows:

$$f \ U \ g = g \ \lor \ (f \ \land \ X(f \ U \ g))$$
$$Fg = g \ \lor \ XFg$$
$$Gf = f \ \land \ XGf$$

At each time step we take the property that must be checked at that time step and rewrite it using the above rules. At the end of the simulation step, we substitute the current signal values in the right hand side (RHS) to determine the temporal property that needs to be checked in the next cycle. We proceed to the next simulation step with this new property.

Example 3.2. Suppose the designer wants to forbid the satisfaction of the property:

$$\varphi \ = \ p \ U \ (q \ U \ r)$$

In other words, the correctness property is $\neg\varphi$, and in order to verify it we search for a match of φ. We start the simulation with the property φ at $t = 0$. Let us assume that the initial signal values are $p = 1, q = 0, r = 0$. At $t = 0$ we rewrite φ as:

$$\varphi = (q \ U \ r) \ \lor \ (p \ \land \ X(p \ U \ (q \ U \ r)))$$
$$= (r \ \lor \ (q \ \land \ X(q \ U \ r))) \ \lor \ (p \ \land \ X(p \ U \ (q \ U \ r)))$$

Substituting the initial values, $p = 1, q = 0, r = 0$, we get:

$$X(p \ U \ (q \ U \ r)) = X\varphi$$

Therefore in the next cycle we need to check φ again. In other words, the checker automaton stays in the start state (see Fig 3.2). We proceed to the next simulation step. Suppose the signal valuations in the next cycle are $p = 0, q = 1, r = 0$. Again we rewrite φ as above and substitute the present state signal values. This time the resulting property for the next cycle is:

$$\psi = q \ U \ r$$

In the next cycle we will rewrite ψ using the same rules:

$$\psi = r \ \lor \ (q \ \land \ X(q \ U \ r))$$

Suppose simulation now returns $p = 1, q = 0, r = 1$. Substituting this in the above formula yields true. Hence we have a match for φ, which is in turn is a refutation of our correctness property. Therefore, we have found a bug! $\square$

Example 3.3. As another example, let us consider the arbiter of Example 3.1 and the property P1, that is:

$$P1: \qquad G[\, r_1 \Rightarrow Xg_1 \wedge XXg_1\,]$$

We will first create the negation of P1 - let us call it φ:

$$\varphi \;=\; F(\, r_1 \wedge (X\neg g_1 \vee XX\neg g_1))$$

At the start state, we need to check φ – we will therefore rewrite it as:

$$\varphi \;=\; (r_1 \wedge (X\neg g_1 \vee XX\neg g_1)) \vee X\varphi$$

As long as the test bench does not assert r_1, the first term in this formula will be false, and therefore our checker automaton will stay at the start state, that is, it will check φ itself in every cycle.

When the test bench drives r_1 (say at time t), the new property will be:

$$(X\neg g_1 \vee XX\neg g_1) \vee X\varphi$$

which is obtained by substituting $r_1 = 1$ in φ. Therefore, in the next cycle, $t + 1$, the checker will look for:

$$\psi = \neg g_1 \vee X\neg g_1 \vee \varphi$$
$$= \neg g_1 \vee X\neg g_1 \vee ((r_1 \wedge (X\neg g_1 \vee XX\neg g_1)) \vee X\varphi)$$

In response to the stimulus $r_1 = 1$, the arbiter implementation of Example 3.1 will assert g_1 at time $t + 1$. Suppose that the test bench drives $r_1 = r_2 = 0$ at time $t + 1$. Substituting this information in ψ gives us:

$$X\neg g_1 \vee X\varphi$$

which yields the property:

$$\eta \;=\; \neg g_1 \vee \varphi$$

for the checker automaton at $t+2$. In the response to the stimulus $r_1 = r_2 = 0$ at time $t + 1$, the arbiter will assert $g_1 = 0$ at time $t + 2$. This will satisfy η and will indicate a match of φ. Since φ is the negation of P1, we have found a refutation of P1, and thereby uncovered a bug! □

Our success in finding the bug of Example 3.3 relies on the assumption that the test bench will drive $r_1 = 1$ at some time t and then drive $r_1 = r_2 = 0$ in the next cycle, $t+1$. If the test bench never drives such a sequence of inputs, the bug will escape detection.

This is one of the major issues in dynamic assertion based verification. On the one hand the industry is moving towards coverage driven randomized test generation to relieve the validation engineer from writing a large number of directed tests. On the other hand assertions often describe temporal properties over complex corner case behaviors that rarely occur in practice, and therefore have a low probability of occurrence in a randomized test generation environment. Writing directed tests for such complex cases is non-trivial. Since the checker automaton actually interprets the property over simulation runs, it makes sense to use the automaton to determine the right kinds of test inputs for the given property. This is one of the main challenges in dynamic ABV. We present some of our research on this topic in Chapter 7.

3.4 Formal Property Verification

Why do we need formal property verification? The most formidable argument in favor of FPV is that the formal properties are checked *exhaustively*, that is, for *all* possible behaviors of the module-under-test under *all* possible input scenarios. The number of possible inputs at a given time t for a module having k inputs is of the order of 2^k. For verifying a *temporal* property we may have to consider *input sequences* over multiple cycles. The number of possible input sequences over n cycles is of the order of n^{2^k}, which grows alarmingly with n and k.

For example, consider the property P4 of Example 3.1:

$$P4: \qquad G[\ \neg r_1 \ \wedge \ X \neg r_1 \ \Rightarrow \ XX \neg g_1\ ]$$

To verify this property, we need to look at input sequences over two cycles. In each cycle, we have $2^2 = 4$ possible input vectors over the two inputs, r_1 and r_2. Therefore we have $2^4 = 16$ input sequences of length 2. If we had an arbiter that arbitrates over 6 request lines, we would need to consider $2^{2^6} = 2^{64}$ input sequences, which is quite formidable.

Therefore the task of verifying properties by exhaustive simulation requires too many test sequences to be feasible in practice. FPV guarantees that the property is automatically verified over *all* such test sequences. If the property is critical for functional correctness, this guarantee is very valuable.

The formal model checking methodology for verifying an LTL property, φ, over an RTL module, M, is intuitively quite simple:

1. We create the checker automaton for $\neg\varphi$.

2. We extract a finite state machine, J, from the module M.

3. We compute the product of J with the checker automaton.

4. If the product contains a run, we have established that the module M has a run which satisfies $\neg\varphi$ (and therefore, refutes φ). We report the existence of a bug, and report the run as the counter-example (or witness).

5. If the product is *empty* (it contains no run), then we report that M satisfies the property, φ.

The main bottleneck of FPV tools is in Step 2. We will discuss this issue shortly.

When a module fails to satisfy a property, the FPV tool will find the bug and produce a counter-example trace. The counter-example establishes the claim of the FPV tool that it has found a bug. On the other hand, if a property holds on a module, then the FPV tool finds out that there is no bug (by checking the emptiness of the product), but has no succinct way to *prove* this result to the validation engineer.

This is a fundamental issue with many known problems. For example, in the Boolean CNF satisfiability problem, it is easy to produce a witness when a given property is satisfiable (any satisfiable assignment to the variables acts as a witness), but there is no succinct proof to show that a property is unsatisfiable.

Therefore the validation engineer has the following problem. If the FPV tool returns a bug, it also produces the counter-example trace as a proof. If the FPV tool returns no bugs, then *we have no easy way to verify whether the tool is right*. We will return to this issue when we dicuss FPV coverage methodologies.

We will now demonstrate the FPV methodology by walking through each of its steps. We will use Example 3.1 as our running example.

3.4.1 Creating the Checker Automaton

In Section 3.1 we outlined the methodology for converting properties into checker automata. There are a few subtle issues that need to be considered while using these automata in FPV. We will study these issues through examples.

Let us consider the property P2 of Example 3.1, namely:

$$P2: \qquad G[\, \neg g_1 \;\Rightarrow\; g_2 \,]$$

The property ensures that the bus is always owned by some master. In order to verify this property over the implementation shown in Example 3.1, we will create the checker automaton for the negation of P2, that is:

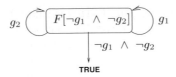

Fig. 3.4. Simplified checker for $F[\neg g_1 \wedge \neg g_2]$

$$\varphi : \qquad F[\; \neg g_1 \wedge \neg g_2 \;]$$

Fig 3.4 shows a simplified VWAA for φ.

How would we verify this property by dynamic verification? Going by the methodology presented in Section 3.3, we will rewrite this property as follows:

$$\varphi : \qquad (\neg g_1 \wedge \neg g_2) \vee X\varphi$$

We will then evaluate the Boolean property:

$$\psi : \qquad \neg g_1 \wedge \neg g_2$$

on the values of g_1 and g_2 in the current simulation cycle. If ψ evaluates to *true* then we report a match for φ. Otherwise, we relegate the remaining term, $X\varphi$, for the next cycle. If ψ never matches up to the end of the simulation we will report that: φ *never matched during simulation, and hence the implementation possibly satisfies P2.*

Dynamic verification checks the properties up to the number of simulation cycles. For FPV we must report the truth of the property considering runs of all lengths. This is made possible by the fact that the implementation has a finite number of states. Therefore every run must eventually loop back to some previous state. The key lies in detecting the loops and interpreting the property appropriately over the loop.

There are broadly two methodologies for checking the properties. These are:

1. *On-the-fly Automata Theoretic Approaches.* This methodology is similar in flavor to the dynamic verification approach. We perform a depth-first traversal over the state machine of the implementation. Along each path generated by the depth-first traversal, we evaluate the properties dynamically. If the path loops back to some previous state we use a set of rules to determine whether the property holds on the path. For example, if we encounter a p-labeled loop while checking for the property Gp, we will accept the loop as one that satisfies Gp. On the other hand, if we encounter a $\neg p$-labeled loop (one in which every state is labeled by $\neg p$) while checking for the property Fp, then we will reject the loop, since it never reaches a p-labeled state.

2. *Tableau based Model Checking.* In this methodology we will construct the checker automaton in a slightly different way, as a closed system. The resulting automaton, called a tableau will contain all paths that satisfy the property. We will take the product of the tableau with the state machine from the implementation and check for emptiness.

We will elaborate the second approach in this section, and present a variation of the first approach while describing SAT-based approaches.

Creation of the Tableau

The tableau is simply a different representation of the checker automaton. It is a state machine, where each state bit is an *elementary subformula* of the given property. Intuitively, these elementary subformulas are of two types - Boolean formulas over the present state variables, and temporal subformulas over future state variables.

The set $EL(f)$ of elementary subformulas of f is recursively defined as follows:

- $EL(p) = \{p\}$ if p is a signal or atomic proposition

- $EL(\neg g) = EL(g)$

- $EL(g \vee h) = EL(g) \cup EL(h)$

- $EL(g \wedge h) = EL(g) \cup EL(h)$

- $EL(Xg) = \{Xg\} \cup EL(g)$

- $EL(g\ U\ h) = \{X[g\ U\ h]\} \cup EL(g) \cup EL(h)$

For example, the set of elementary subformulas of the property:

$$\varphi : \qquad F[\ \neg g_1\ \wedge\ \neg g_2\ ]$$

consists of g_1, g_2, $X\varphi$. These three elementary subformulas represent the state bits of the tableau, and thereby yields $2^3 = 8$ states.

The elementary subformulas are really the ones whose truth dictates the state of the checker. For example, the property φ above can be true at a state in the following two ways:

1. *Both g_1 and g_2 are false at that state.* The truth of $X\varphi$ is irrelevant here – hence these cases are represented by two states of the tableau, namely: $\{\neg g_1, \neg g_2, X\varphi\}$ and $\{\neg g_1, \neg g_2, \neg X\varphi\}$.

2. g_1 or g_2 is true at that state, and $X\varphi$ is also true at that state. These cases are represented by three states of the tableau, namely: $\{g_1, g_2, X\varphi\}$, $\{g_1, \neg g_2, X\varphi\}$ and $\{\neg g_1, g_2, X\varphi\}$.

The second case needs more attention. $X\varphi$ is true at a state only if φ is true at the next state. Therefore, the tableau must have a transition from a state satisfying $X\varphi$ to a state satisfying φ. Similarly the tableau must have transitions from states satisfying $X\neg\varphi$ to states satisfying $\neg\varphi$. Fig 3.5 shows the resulting tableau.

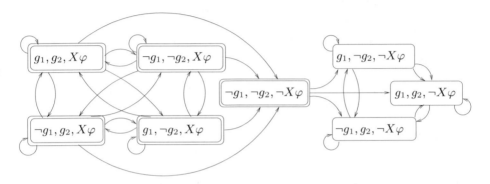

Fig. 3.5. Tableau for $F[\neg g_1 \wedge \neg g_2]$

Let us refer to runs that satisfy φ as φ-satisfying runs. It is easy to see that every φ-satisfying run belongs to the tableau and that these runs start from the states which are either labeled by $\neg g_1 \wedge \neg g_2$, or are labeled by $X\varphi$. We shall refer to these states (shown with double borders in Fig 3.5) as φ-satisfying states. Every run that starts from the rest of the states gets trapped between the three states $(\neg g_1, g_2, \neg X\varphi)$, $(g_1, \neg g_2, \neg X\varphi)$, $(g_1, g_2, \neg X\varphi)$, and therefore does not satisfy φ.

Is every run starting from a φ-satisfying state a φ-satisfying run? Unfortunately no. Consider the runs that remain forever within the three states, $(\neg g_1, g_2, X\varphi)$, $(g_1, \neg g_2, X\varphi)$, and $(g_1, g_2, X\varphi)$. These runs do not eventually reach any state that satisfies $\neg g_1 \wedge \neg g_2$, and are actually not φ-satisfying runs. Therefore only those runs are φ-satisfying runs, that start from φ-satisfying states but do not get trapped in $g_1 \vee g_2$ satisfying states. In general for any property of the form, $f \ U \ g$, the run must not get trapped in $\neg g$ labeled states. This is a typical instance of a *fairness constraint*.

As per our verification strategy, we seek a φ-satisfying run in the implementation. This may be done by computing the product of the tableau with the state machine of the implementation and then checking whether the product has any fair run starting from a φ-satisfying state.

3.4.2 FSM Extraction

Since FPV must necessarily consider every possible run of the implementation, FPV techniques require the state machine model of the implementation. Building the state machine model of the implementation is the single most complex step of FPV, and accounts for the capacity limitations of existing tools.

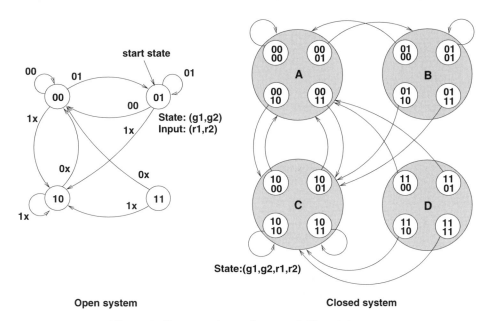

Fig. 3.6. State machine: Open and Closed forms

Fig 3.1 shows the state machine of our running example – the arbiter. Since the temporal properties do not distinguish between input and output signals, we must transform this state machine into a *closed system* – where inputs are also part of the state definition. This is a simple transformation, and the new state machine model is shown besides the original one in Fig 3.6. Each state of the original machine is now a cluster of four states (shown as A, B, C, and D) – the four states in a cluster differ only in the input bits, that is, they have the same values for g_1 and g_2. Each transition of the original machine is represented by four transitions. To save space, Fig 3.6 shows these edges as single edges leading to a block of states. For example, the transition from state 01 to state 00 on input 00 in the original state machine is now represented by four non-deterministic transitions, namely (0100, 0000), (0100, 0001), (0100, 0010) and (0100, 0011). We show these four transitions in Fig 3.6 by a single edge from the state 0100 to the block A, which consists of the four possible next states. The new state machine is non-deterministic, simply because the next state inputs are not known a priori.

It is interesting to note that the complexity of both CTL and LTL model checking is linear in the size of the state machine of the implementation. Intuitively this means that once we are able to build the state machine model of an implementation, the task of checking the properties on the state machine is not too complex.

Synthesis tools have been synthesizing the RTL code of sequential circuits into hardware for several decades. Sequential circuits are state machines. *If the synthesis tool did not have any problem in creating the state machine from the RTL, why do FPV tools run into capacity problems while extracting a state machine model?*

The answer lies in concurrency and compositionality. When we have two or more sequential components in the RTL, we can synthesize each component as a separate state machine. In hardware, these state machines will run in parallel (possibly with the same clock). On the other hand, if we are given a formal property over the signals of one or more sequential components, then FPV requires the *global state machine* obtained by computing the product of the state machines of the components. This product leads to state explosion.

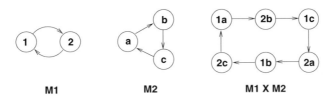

Fig. 3.7. Product of state machines

Fig 3.7 shows the product of two state machines, $\mathcal{M}_1$ and $\mathcal{M}_2$. In general if a module M has k component modules, $M_1, \ldots, M_k$, having respectively $n_1, \ldots, n_k$ states, then M has at least $n_1 \times n_2 \times \ldots \times n_k$ states. This figure grows alarmingly with increase in design complexity, and is popularly known as *state explosion*.

How do we contain state explosion in FPV? This is the most burning question within the FPV community. Not surprisingly there are many competitive answers. These include:

1. *Compact representations of state machines.* Over the last decade, several interesting ways of representing state machines have been proposed. Two representations stand out among these in terms of popularity and actual use in practice. These are *Binary Decision Diagrams* (BDD) and *Boolean*

SAT formulas. The FPV approach with these representations has been outlined in Section 3.5 and Section 3.6 respectively.

2. *Compositional verification.* Compositional verification actually refers to the task of performing FPV *without* computing the composition of the component modules! One of the most popular approaches in this area is that of *assume-guarantee* verification, where each component module is verified in isolation under specific assumptions about the behavior of its environment. The assumptions made while verifying a component must be guaranteed by the rest of the components.

3. *Approximate verification.* In approximate verification, we simply remove some of the components. If these components drive some of the remaining components, we treat those signals as primary inputs into the remaining components. If the property passes under this restriction, we are able to guarantee that it holds on the original design. On the other hand, if the property fails then it may not actually be a bug, since the counter-example scenario may not be driven by the removed components. However since FPV tools always produce a counter-example on failure, the validation engineer can easily verify whether the bug is real by simulating the counterexample trace. Therefore, this is a safe approach.

4. *Design intent coverage.* This is a new paradigm for FPV recently proposed by our research group. In this approach, the verification task is manually decomposed, but the soundness and completeness of the decomposition is automatically verified. We believe that this is a very practical way to handle designs where existing FPV tools run into capacity problems. We present the details of this approach in Chapter 6.

Doing justice to each of the above approaches is beyond the scope of this book. The focus of this book is inclined towards the validation engineer who intends to use a given FPV tool – hence we will only outline those core approaches that are part of most existing industry FPV tools.

3.4.3 Computing the Product

The final step of our verification strategy is to compute the product of state machine of the implementation with the tableau of the negation, φ, of the original formal property, and check whether the product contains a fair run starting from a φ-satisfying state of the tableau.

Fig 3.8 shows a fair run that is common between the state machine model of the implementation shown in Fig 3.6 with the tableau shown in Fig 3.5. This is one of many runs that may appear in the product of the two machines – each such run serves as a counter-example for the property P2 of Example 3.1.

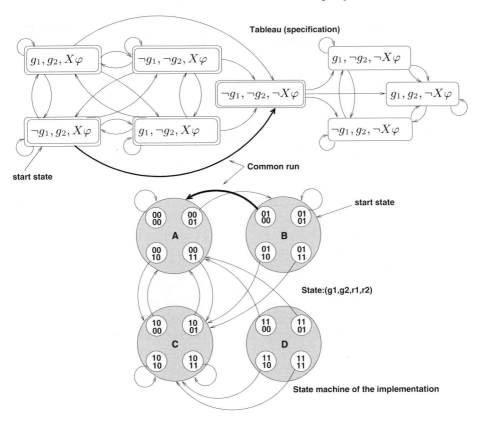

Fig. 3.8. Tableau vs implementation

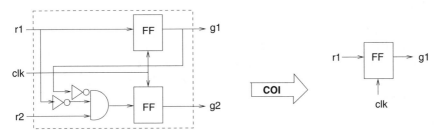

Fig. 3.9. Cone of influence reduction with $G[\, r_1 \Rightarrow Xg_1 \ \wedge \ XXg_1 \,]$

In reality the product computation does not necessarily begin *after* the first two steps. Keeping in mind that the product computation is our end-goal, FPV tools use the properties to perform several standard optimizations upfront, in order to contain the size of the state machine of the implementation. We will discuss three of the most widely adopted optimizations here:

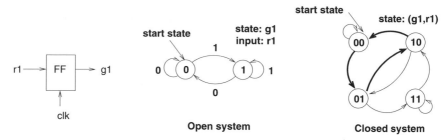

Fig. 3.10. Reduced state machine, after COI

1. *Cone-of-influence reductions.* The idea here is to prune the circuit to re-move all components that do not (directly or indirectly) influence the truth of the given property. For example, in order to verify the property:

$$\text{P1:} \quad G[\, r_1 \Rightarrow Xg_1 \wedge XXg_1 \,]$$

on the implementation of Fig 3.1, we observe that the logic for the output g_2 does not influence the truth of this property. We can therefore prune the circuit as shown in Fig 3.9. The reduced state machine is much smaller, as shown in Fig 3.10, which contains only those paths that are relevant to the final product. In order to check P1, we create its negation:

$$\psi = F[\, r_1 \wedge (X\neg g_1 \vee XX\neg g_1)]$$

The bold edges in the closed system representation shown in Fig 3.10 demonstrates a run that satisfies this property and thereby serves as a counter-example for P1 in the implementation.

Cone-of-influence reductions typically add a lot of value to an FPV tool, and these are an integral part of most FPV tools. Therefore if the designer can express the formal specification in terms of small properties over a few variables, then typically cone-of-influence reductions are able to effect sufficient pruning to avoid state explosion.

2. *Pruning using "assume" constraints.* Most property specification lan-guages will allow the engineer to specify *assert* statements to describe the correctness requirements and *assume* constraints to indicate the as-sumptions about the inputs, under which the assertions are expected to hold.

Intuitively, *assume constraints* can be used to prune the state machine of the implementation, since we need not consider those runs of the imple-mentation that violate the assume constraints. In many cases, this can lead to significant reduction in the size of the reduced state machine.

Practioners of FPV have also reported an exactly opposite experience at times with their FPV tools. They have observed that in some cases adding assume constraints into the specification have aggravated the capacity problem!

This is *not* surprising. The state machine of the implementation may have to grow in size in order to incorporate a restriction which is temporal in nature. For example, consider the simple closed system state machine shown in Fig 3.10. Suppose we add the assumption:

$$G[\ \neg r_1 \ \lor \ \neg X r1 \ \lor \ \neg XX r_1]$$

which says that r_1 *never comes in three consecutive cycles.* In order to incorporate this information into the state machine of Fig 3.10, each state must *remember* the status of the request line in the previous cycle. This will lead to an increase in the number of state bits, and thereby the number of states of the state machine.

3. *Simulation equivalence relations.* The state machine of an implementation may have more details than are required for verifying a given property. For example, consider the property P2 of Example 3.1:

$$P2: \qquad G[\ \neg g_1 \ \Rightarrow \ g_2\ ]$$

In order to verify this property we look for a path satisfying its negation:

$$\varphi = F[\ \neg g_1 \ \land \ \neg g_2\ ]$$

In order to check this property, we need to check whether any state satisfying $\psi = \neg g_1 \ \land \ \neg g_2$, is reachable from the start state of the implementation. In other words, we may partition the states into two parts, those that satisfy ψ, and the rest. We want to determine whether it is possible to reach the first partition from the start state.

We can minimize the state machine based on such partitions. For example, the reduced state machine after partitioning the state machine of Fig 3.6 around ψ is shown in Fig 3.11. The reduced state machine has only two states as compared to sixteen states of the original state machine.

There are standard algorithms for minimizing a state machine based on a well defined equivalence relation. In the context of model checking, two of the most popular forms of equivalence are *bisimulation equivalence* and *stuttering equivalence.* Stuttering equivalence preserves the truth of untimed properties, such as Fq, where the number of states on a path leading to a q-labeled state is not important – these states may therefore be condensed into one state. On the other hand, in order to check a timed temporal property, such as, $F_{[3,9]}\ q$, we need to count the number of states on the path leading to a q-labelled state, and check whether this number

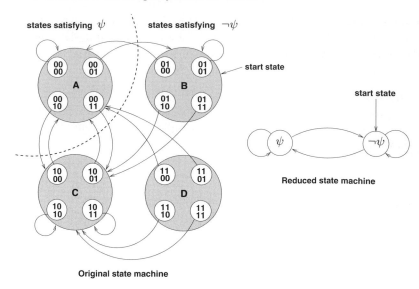

Fig. 3.11. Equivalence with respect to $\psi = \neg g_1 \wedge \neg g_2$

is between 3 and 9. Therefore we cannot merge these states. However, two paths reaching the same q-labelled state in the same number of steps can be condensed into one path. Bisimulation equivalence is timing preserving.

3.5 BDD-based Formal Property Verification

Binary Decision Diagrams (BDDs) are compact canonical representations of Boolean functions. Decision trees based on Shannon's expansion have been used for propositional reasoning in Artificial Intelligence for many years. BDDs utilize self-similarity in such structures to give a more compact representation than decision trees.

The use of BDDs in FPV was instrumental in bringing the technology into practice. Initial results show that the BDD-based FPV techniques were able to handle as much as 10^{20} states – which was unthinkable with explicit state space search. BDD-based FPV tools have several characteristic features, and several tunable parameters. Our goal will be to understand the meaning of these features so that the behavior of BDD-based FPV tools, and particularly their limitations become apparent to us.

In this section we will study the structure of BDDs and attempt to understand their benefits and limitations in the context of formal property verification. Formally, any model checking technique that works on a symbolic representation of the implementation can be called *symbolic* model checking. However the term *symbolic model checking* is sometimes popularly interpreted as BDD-based model checking.

3.5.1 What is a BDD?

Given a Boolean function $f(x_1, \ldots, x_k)$ over variables, $x_1, \ldots, x_k$, we can rewrite f as:

$$f(x_1, \ldots, x_k) = (x_1 \wedge f_{x_1 \leftarrow 1}) \vee (\neg x_1 \wedge f_{x_1 \leftarrow 0})$$

where $f_{x_1 \leftarrow 1}$ denotes f with x_1 substituted by 1, and $f_{x_1 \leftarrow 0}$ denotes f with x_1 substituted by 0. This is known as *Shannon's decomposition* of f on the variable x_1.

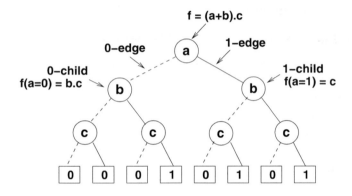

Fig. 3.12. Decision tree for $f = c \wedge (a \vee b)$

Fig 3.12 shows a decision tree for the Boolean function:

$$f(a, b, c) = c \wedge (a \vee b)$$

Each node of the tree represents a Boolean function realized by the subtree rooted at that node, and is labeled by a variable which is used for Shannon's decomposition to determine the functions represented by its children. For example if a node represents the function $f(x_1, \ldots, x_k)$, and if the node is labeled by x_i, then the node has two children – the 0-child, representing $f_{x_1 \leftarrow 0}$, and the 1-child representing $f_{x_1 \leftarrow 1}$. The edge leading to the 0-child is called the 0-edge, while the edge leading to the 1-child is called the 1-edge. We show the 0-edges with dashed lines in Fig 3.12.

In order to compute the function value for a given valuation of the variables, we can traverse the tree as follows. We start from the root, and follow either the 0-edge or the 1-edge depending on the value of the variable at the root. We continue this process until we reach a leaf node. The value of the leaf node gives us the value of the function for the given valuation of the variables.

The sequence in which we choose the variables for Shannon's decomposition is referred to as the *variable order*. In Fig 3.12, the variable order is $a \prec b \prec c$. The decision tree is *ordered* if the variable order is the same on all paths of the tree.

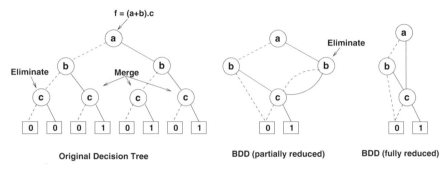

Fig. 3.13. BDD for $f = c \wedge (a \vee b)$

Several nodes of the decision tree may represent the same Boolean function. BDDs make good use of this fact by *sharing* such nodes. The existence of a variable order guarantees that the resulting digraph is acyclic. Fig 3.13 shows the steps in reducing the decision tree of Fig 3.12 to a BDD using self-similarity. This has been done using the following two rules:

1. *If two nodes represent the same function, then we merge them.* Fig 3.13 shows the merging of three c-labeled nodes.

2. *If a node has the same 0-child and 1-child, then that node represents a "don't care" variable, and is removed.* Formally, it follows from Shannon's decomposition that f is independent of x_i whenever $f_{x_i \leftarrow 0} = f_{x_i \leftarrow 1}$. This explains the elimination of the node indicated in Fig 3.13.

Reduced ordered Binary Decision Diagrams (ROBDD) are *canonical* in nature. This means that if f and g are two representations of the same Boolean function, then they will have the same ROBDD. Henceforth we will loosely use the term, BDD, to mean a ROBDD.

Canonicity is a very useful property for formal equivalence checking. If we wish to verify whether a given RTL module is an adder, we can extract its logic and create a BDD. We can also create the BDD for the addition function (which is a Boolean function). If the two BDDs are identical, then the RTL correctly implements an adder. Otherwise, the RTL has a bug.

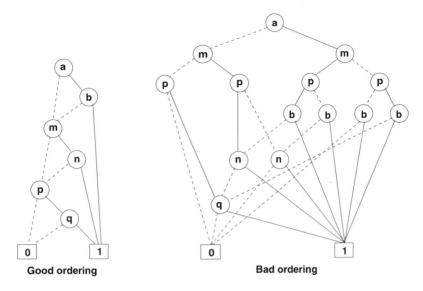

Good ordering   **Bad ordering**

Fig. 3.14. Effect of variable ordering on BDD size

Not surprisingly, BDDs are widely used for formal equivalence checking. The main limitation here is that the size of a BDD can grow rapidly with the number of variables. One of the parameters that determine the size of a BDD is the variable ordering. Fig 3.14 shows two BDDs for the function:

$$f = (a \wedge b) \vee (m \wedge n) \vee (p \wedge q)$$

The first BDD uses the ordering, $a \prec b \prec m \prec n \prec p \prec q$, while the second BDD uses the ordering, $a \prec m \prec p \prec b \prec n \prec q$. The first ordering is clearly better than the second.

Theoretical lowerbounds show that the task of finding the optimal variable ordering for a function is a hard problem. Worse still, it has been shown that the task of improving the variable ordering is also hard in general. However, BDD packages support a wide variety of heuristics for finding a *good* variable ordering. Though familiarity with these heuristics can enable a validation engineer to tune the performance of the FPV tool to a large extent, it is a formidable challenge to master and interpret the benefits of these heuristics in a given context. Most existing tools therefore handle such choices internally. There are two notable issues here:

1. *Dynamic Variable Ordering.* When we build the BDD for a complex circuit, we start with the BDDs for the smaller sub-circuits and then use logical operations on pairs of BDDs to develop the BDDs for the circuits which contain the smaller sub-circuits. For example, Fig 3.15 shows a circuit and the sequence in which we build the BDD for the circuit. BDD

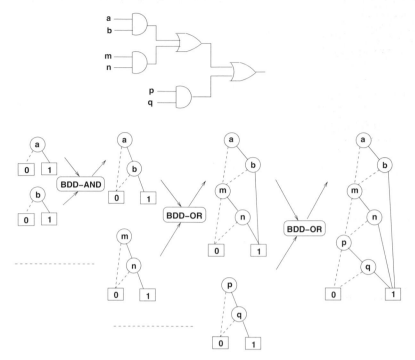

Fig. 3.15. Incremental construction of BDDs for a circuit

packages support this approach by providing functions for performing all Boolean operations (NOT, AND, OR, etc) over BDD representations of Boolean functions.

The main problem here is that the best variable ordering for the sub-circuits may not be the best one for the whole circuit. Therefore, as we begin to build the BDD of the circuit from the BDDs of its components, the ordering has to be dynamically changed to keep the BDD sizes small.

Dynamic variable ordering is a nice variable re-ordering heuristic which strives to reduce the BDD size by making local changes in the ordering. For circuits of small size this is an overhead. While building BDDs for large circuits incrementally, this heuristic works remarkably well, and often makes the difference between feasibility and infeasibility. Unlike the static ordering heuristics, dynamic variable ordering works in the background and works all the time.

Many BDD-based FPV tools have a parameter that allows the validation engineer to set dynamic variable reordering ON/OFF. It is an interesting exercise for the new practioner of FPV to experiment with this parameter.

2. *There are circuits having no good ordering.* The most commonly cited case is that of a multiplier. The multiplication function always tends to grow in the number of terms with Shannon's decomposition. It has been shown that no variable ordering can contain the growth of the BDD of a multiplier with the number of bits of its operands.

To get around this problem several alternative representations have been proposed for arithmetic circuits. These include *Binary Moment Diagrams* (BMD), *Arithmetic Decision Diagrams* (ADD) and *Mutli-Terminal BDDs* (MTBDD). Most of the public domain decision diagram packages support all of these representations. Some of the internal FPV tools used in specific companies allow the validation engineer to choose the type of representation. This choice is likely to become visible in future generation FPV tools as well.

BDDs have been used in a wide range of CAD problems, including equivalence checking, symbolic simulation, symbolic model checking, state minimization, and false path identification. Several public domain BDD packages are available. These include, CUDD, TUDD, BUDDY and CMUBDD. Information about BDD packages can be found in the website: http://www.bdd-portal.org.

3.5.2 BDDs for State Machines

Sequential circuits may typically be viewed as a collection of sequential elements (such as flip-flops) and combinational logic. In Fig 3.15 we demonstrated the methodology for creating a BDD from a given combinational circuit. Intuitively, we may use the following steps to create a BDD representation for a sequential circuit.

1. Drop the sequential elements (say flip-flops) from the circuit to extract the combinational logic.

2. For each flip-flop do the following:

   a) Treat its output, y, as a primary input to the combinational logic. y represents a *present state bit* of the state transition relation of the sequential circuit.

   b) Treat its input as a primary output of the combinational logic. Call this output y'. y' represents the *next state variable* for y in the state transition relation of the sequential circuit.

3. Build the BDD for the modified combinational logic.

The BDD for the combinational logic actually represents the *state transition relation* of the circuit. Fig 3.16 shows the construction for the circuit of Example 3.1. We have one BDD for each next state bit – each BDD expresses the value of a next state bit as a function of the present state bits and input bits.

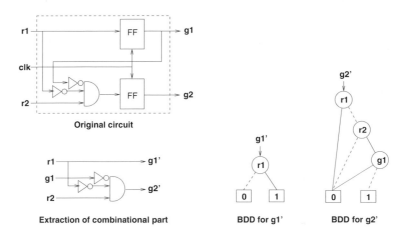

Fig. 3.16. BDDs for transition functions

The BDD for each next state bit represents a component of the state machine of the circuit. In order to find the next state for a given state under a given input, we can traverse each BDD individually to determine the value of each next state bit. However, for FPV we need to consider the global state machine created by taking the product of the components. This is the main source of state explosion. We will present an outline of the methodology in the next section.

3.5.3 Symbolic Reachability

Given the BDDs for each component state machine, how can we determine whether a given state is reachable from the start state? This is the core task for every BDD-based FPV tool.

Let us consider the property, P3, of Example 3.1:

$$P3: \qquad G[\, \neg g_1 \, \vee \, \neg g_2 \,]$$

In order to check this property, we need to determine whether any state satisfying $g_1 \wedge g_2$ is reachable from the start state.

It is not possible to answer this question by examining each component of Fig 3.16 in isolation. For example, the BDD for g_1' shows that whenever r_1

is high, g_1 will be high in the next cycle. The BDD for g_2' shows that g_2 will be high in the next cycle when r_1 and g_1 are low and r_2 is high. Therefore individually g_1 can be high in future, and so can g_2. *Can they be high together?* We need to look at the global state machine to answer this question.

BDD for the Global State Machine

We will use the notion of *characteristic functions* to build a BDD for the global state machine. The idea is simple. Suppose we have k state bits. We therefore have $n = 2^k$ states. A total of n^2 transitions are possible between these n states. The state transition relation specifies the set of transitions that actually belong to the state machine of the circuit. The characteristic function, f, of the state transition relation is a Boolean function over the present state bits, input bits, and next state bits, which returns true for valid transitions (those which belong to the state machine) and false for the rest.

The characteristic function, f, for the state machine of Example 3.1 is shown in Fig 3.17. We use the symbol, x, to denote a *don't care*. The first four rows cover all the valid state transitions. The value of f is true on these vectors. For the rest of the vectors, f is false.

$g_1 g_2$ $r_1 r_2$	$g_1' g_2'$	f
xx 00	00	1
0x 01	01	1
1x 01	00	1
xx 1x	10	1
All other 6-bit vectors		0

Fig. 3.17. Characteristic function for the arbiter state machine

We can build the BDD for the characteristic function of the state machine directly from the BDDs for the individual next state bits. By definition, the characteristic function for a given state bit has the same inputs as the logic cone driving that state bit, plus the next value of the state bit itself. Thus it is equivalent to the expression: *logic cone = next value*, which can be computed by taking the EXNOR (which represents the equality function) of the BDD for the logic cone and the BDD for the next value.

Fig 3.18 shows the steps. We first build the characteristic function BDDs (CF-BDDs) for each component by taking the exclusive-NOR (EXNOR) of the component BDD with the next state variable. For example to build the CF-BDD for the next state bit g_1', we take the EXNOR of the component BDD with the variable, g_1'.

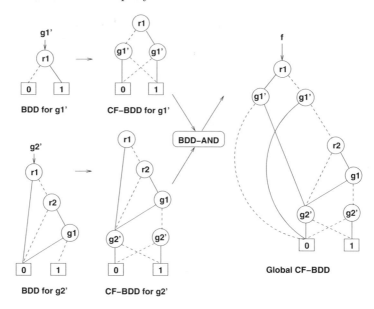

Fig. 3.18. BDD for state transition relation

To build the BDD for the global transition relation, we simply take the conjunction (AND) of the component CF-BDDs as shown in Fig 3.18. It may be noted that every path to the 1-node (*true* node) in the final BDD represents a valid transition of the state machine of Example 3.1.

Taking the conjunction of the CF-BDDs is an expensive step in practice. This typically leads to a blow-up in the size of the BDD. Therefore symbolic reachability algorithms often use *partitioned transition relations*, where the objective is to avoid taking the conjunction of BDDs with large number of variables. More details of this approach are given in [38].

Symbolic State Traversal

In order to verify the property, P3, of Example 3.1 we need to determine whether any run from the start state leads us to a state satisfying $g_1 \wedge g_2$. Let us see how this search may be performed using the BDD of Fig 3.18.

Suppose the start state for the arbiter of Example 3.1 is the state, $g_1 = 0, g_2 = 1$. Fig 3.6 shows that this represents a set S_0 of four states over (g_1, g_2, r_1, r_2), namely:

$$S_0 = \{\ 0100,\ 0101,\ 0110,\ 0111\ \}$$

Our approach will intuitively be as follows. We will begin from the start states, S_0, and check whether it contains any state satisfying $g_1 \wedge g_2$. If not, we will find the set, S_1, of states reachable from the states in S_0 in at most one cycle. If S_1 also does not contain any state satisfying $g_1 \wedge g_2$, then we will find the set, S_2 of states reachable from the states in S_0 in at most two cycles, and look for a state satisfying $g_1 \wedge g_2$ in S_2. *How long will we continue in this manner?*

Off course, if some state satisfying $g_1 \wedge g_2$ is reachable in k-cycles, then we will terminate after at most k-iterations by finding one such state. *What happens if there is no reachable state satisfying $g_1 \wedge g_2$?*

Since our state machine has a finite number of states, each reachable state can be reached in a finite number of cycles. Therefore, if we continue the generation of the state sets, $S_0, S_1, \ldots$, we will find that after a finite number of iterations, $S_i = S_{i+1}$. In other words, some S_i will contain all reachable states, and thereafter, the state sets, $S_{i+1}, S_{i+2}, \ldots$, will all be equal to S_i. Therefore we can terminate whenever we find $S_i = S_{i+1}$. If S_i does not contain any state satisfying $g_1 \wedge g_2$ at that point, then we can conclude that no state satisfying $g_1 \wedge g_2$ is reachable from a start state. The set S_i is called the *fixed point* for the reachability function.

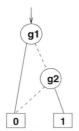

Fig. 3.19. BDD for start state S0

We will use BDDs for representing the state sets, $S_0, S_1, \ldots$, succinctly. Fig 3.19 shows the BDD for S_0. Fig 3.20 shows the first iteration – computing S_1 from S_0 using the BDD for the transition relation (Fig 3.18). The steps are as follows:

1. A transition is a tuple $\langle g_1, g_2, r_1, r_2, g_1', g_2' \rangle$, where $g_1 g_2$ is the present state, $r_1 r_2$ is the input and $g_1' g_2'$ is the next state. Recall that the BDD, R, for the transition relation is a collection of valid transitions – every valid transition is a path to the 1-labelled node in this BDD. The AND operation on the BDDs, S0 and R, gives us the subset of valid transitions for which the present state, $g_1 g_2$, belongs to S_0.

2. If we look at the next state bits, $g'_1 g'_2$, in those transitions where the present state, $g_1 g_2$, belongs to S_0, then we will get exactly the set of next states for the states in S_0. Therefore we existentially eliminate[1] all other variables except the next state bits, g'_1 and g'_2 to obtain the BDD called NS0-Temp in Fig 3.20.

3. We rename g'_1 to g_1 and g'_2 to g_2 in NS0-Temp. This gives us the BDD, NS0, which represents the set of next states for the set of states in S_0. In other words these are the states that can be reached in one cycle from the set of states in S_0.

4. The desired set of states, S_1, is computed by taking the union (logical OR) of the BDDs, NS0 and S_0. The BDD, S1, represents the set of states reachable in 0 or 1 cycles.

Is there any state satisfying $g_1 \wedge g_2$ *in* S_1? The answer is no, since substituting $g_1 = g_2 = 1$ in the BDD, S1, leads us to 0. Therefore we conclude that no state satisfying $g_1 \wedge g_2$ can be reached in 0 or 1 cycles. We do not know yet whether any state satisfying $g_1 \wedge g_2$ can at all be reached.

Therefore, we apply the same steps again to compute S_2 from S_1. The BDD for S_2 will represent all states that can be reached by 0, 1, or 2 cycles. In this case we will find that $S_2 = S_1$. Therefore S_1 contains all states that are reachable. Since this set does not contain any state satisfying $g_1 \wedge g_2$, we can safely conclude that no such state is reachable. This is turn proves that the mutual exclusion property:

$$G[\ \neg g_1 \ \vee \ \neg g_2\ ]$$

is satisfied by our arbiter implementation.

As another example, suppose we intend to verify the property:

$$P2: \qquad G[\ \neg g_1 \ \Rightarrow \ g_2\ ]$$

In this case, we will look for the reachability of a state satisfying $\neg g_1 \ \wedge \ \neg g_2$. This is not satisfied by S_0, hence we compute S_1 from S_0. We find that S_1 contains states satisfying $\neg g_1 \ \wedge \ \neg g_2$, since setting $g_1 = g_2 = 0$ leads us to 1 in the BDD, S1. Therefore, without looking any further, we may conclude that the property P2 fails in our arbiter implementation.

What is the advantage of using BDDs for reachability analysis? The main advatantage is that BDD operations allow us to work on sets of states at a time. For example, in order to find the set of states, S_k (that is, states reachable by k cycles) from the set of states, S_{k-1}, we need to perform only a handful of BDD operations – as shown in Fig 3.20. Explicit traversal of

---

[1] Existential elimination of variable x from function f yields the function $f_{x \leftarrow 0} \vee f_{x \leftarrow 1}$

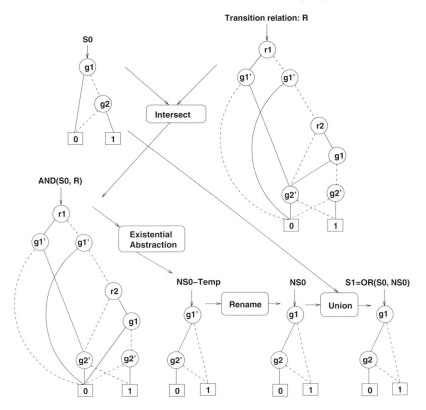

Fig. 3.20. One step of forward reachability

all k-length paths in the state machine would clearly be infeasible when the number of states is vast.

The reachability analysis shown here is known as *forward reachability*. In order to check whether a state y can be reached from a state x, *forward reachability* methods start from x and work forward till we reach y or the fixed point. On the other hand, there are *backward reachability* methods that start from y and work backwards till we reach x or the fixed point.

The steps for *backward reachability* are very similar to those for forward reachability. S_0 denotes the set of target states. S_i denotes the set of states from which one or more states of S_0 can be reached in i steps. We compute S_i from S_{i-1} by taking the union of S_{i-1} with the set of parents of the states in S_{i-1}. In order to compute the parents of the states in S_{i-1}, we do the following:

1. We rename the state variables in S_{i-1} with the corresponding next-state variables. For example, g_1 will be renamed as g_1'.

2. We take the product (logical AND) of S_{i-1} with the transition relation, R. This gives us all transitions where the child belongs to S_{i-1}. We existentially eliminate all variables except the present state bits (say, g_1, g_2) to get the parents of the states in S_{i-1}.

Backward reachability will also reach a fixed point since every run, when traced backwards, will revisit some state in a finite state machine.

BDDs can be used for reachability analysis only when the BDDs can fit in memory. Unfortunately, the size of the BDD for the transition relation tends to grow very fast with the number of state and input variables. As a result BDD-based tools have serious capacity limitations. In spite of this limitation, BDD-based FPV tools add considerable value to the design validation flow for verifying small but complex modules.

BDD-based FPV tools automatically create these BDDs from the design implementation, and automatically apply these symbolic reachability methods to verify properties. In the next section we demonstrate how symbolic reachability can be used to verify CTL properties.

3.5.4 CTL Model Checking

Verification of CTL properties works with simple symbolic reachability methods. Let us start with simple properties.

- The BDD for any Boolean function over the state and input variables actually represents the set of states satisfying that function. In order to test whether a given state satisfies that function, we may substitute the state bits into the BDD and check whether it returns true or false.

- $\neg f$. We first compute the BDD for f and then take its complement (using the BDD-NOT operation).

- $EX\ f$. We first compute the BDD for f – let us call it Z. We then use one step of *backward reachability* to get the set of states satisfying $EX\ f$.

- $E[\ f\ U\ g\ ]$. We can characterize the states that satisfy $E[\ f\ U\ g\ ]$, as follows:

  1. States that satisfy g, and

  2. States that satisfy f and have a next state satisfying $E[\ f\ U\ g\ ]$.

We start with the states that satisfy g and work backwards along f-satisfying states until we can add no more new states. In other words, we use the recursive formulation:

$$E[\ f\ U\ g\ ]\ =\ g\ \lor\ (\ f\ \land\ EX(E[\ f\ U\ g\ ])\,)$$

For example, let us call the BDD for the set of states satisfying g as Z_0. We then use one step of backward reachability to compute, $EX(Z_0)$, which is the BDD for the set of parents of the states in Z_0. The intersection of $EX(Z_0)$ with the BDD for states satisfying f gives us the BDD for the set of states that satisfy f and have a next state satisfying g. The union of this BDD with Z_0 gives us the set, Z_1, of states that satisfy $E[\ f\ U\ g\ ]$ in one or two steps.

In the same way, to compute Z_i we use one step of backward reachability from the BDD, Z_{i-1}, intersect the resulting BDD with the BDD for f, and take the union with Z_{i-1}. We repeat these steps until we reach the fixed point, $Z_k = Z_{k-1}$. At this point Z_k contains all states that satisfy $E[\ f\ U\ g\ ]$.

The functions for the above basic formulas can be used to find states satisfying other formulas as well. We use the following reductions to express other formulas in terms of these basic formulas:

$$
\begin{aligned}
AX\ f &\equiv \neg EX\ \neg f \\
EF\ f &\equiv E[\ true\ U\ f\ ] \\
AF\ f &\equiv \neg EG\ \neg f \\
AG\ f &\equiv \neg EF\ \neg f \\
A[\ f\ U\ g\ ] &\equiv \neg[\ (EG\neg g)\ \vee\ (E[\ \neg g\ U\ (\neg g\ \wedge\ \neg f)\ ])\ ]
\end{aligned}
$$

The last reduction follows from the fact that $A[\ f\ U\ g\ ]$ is not true at a state, s, if and only if at least one of the following applies:

1. g is never satisfied on some run from s. In this case, s satisfies $EG\neg g$.

2. There exists a run starting from s, where f becomes false before g becomes true. In this case, s satisfies $E[\ \neg g\ U\ (\neg g\ \wedge\ \neg f)\ ]$.

In all of the above, if f, g, are also temporal properties, then we recursively compute the sets of states satisfying f, g, and then use the BDDs for those sets while treating formulas containing f, g.

BDD-based LTL model checking is performed by transforming LTL model checking into CTL model checking as follows:

1. The tableau is represented as a BDD, Z. Thus, Z contains all valid transitions present in the tableau.

2. We take the product of Z with the BDD for the transition relation of the implementation. Let the BDD for the product be T.

3. Recall that the last step of the LTL model checking strategy is to check whether the product of the tableau and the implementation is empty,

that is, whether the product has any fair path. This check is performed by verifying the CTL formula $EG[\ true\ ]$ on T under fairness constraints.

Details on CTL model checking under fairness constraints can be found in Clarke, Grumberg and Peled [38].

3.6 SAT-based Formal Property Verification

BDD-based FPV tools typically run into capacity problems because the BDDs tend to grow very fast with the number of state variables. Various tricks have been attempted to keep the BDD sizes within feasible limits, but beyond a point none of these tricks are adequate to control the growth and the FPV tool becomes ineffective.

BDDs are canonical representations of Boolean functions, and they are more compact than the truth table representations, but they are typically much larger than the size of the circuit representation (say in Verilog). A high level description of the circuit is a collection of Boolean equations. A sequential circuit description is also a collection of Boolean equations which describes the transition relation of the sequential circuit.

For example, our arbiter circuit of Fig 1.1 can be succinctly described by the Boolean functions:

$$(r_2 \land \neg r_1 \land \neg g_1) \Rightarrow g_2'$$
$$r_1 \Rightarrow g_1'$$

These Boolean functions are smaller than the BDD for the transition relation shown in Fig 3.18. In general the difference between the sizes of these two representations is significant.

Why then did we use BDDs? Because it enabled us to perform symbolic reachability analysis without *explicitly* traversing all paths of the implementation state machine. Symbolic reachability is the heart of FPV approaches, since explicit state traversal is clearly unfeasible in practice.

Can we perform symbolic reachability without using BDDs? The answer is yes, and the alternative methodology is called *SAT-based reachability*.

3.6.1 What is SAT?

SAT is the traditional short form for the Boolean satisfiability problem. Given a Boolean formula, f, we are required to determine whether f is satisfiable, that is, whether there exists any valuation of the variables in f, under which f evaluates to TRUE. Boolean satisfiability is known to be a hard problem in general – nonetheless, we use SAT because it works well enough in practice.

For example, the following formula is satisfiable for all valuations of a and b where $a \neq b$.

$$f = (a \lor b) \land (\neg a \lor \neg b)$$

The following formula is unsatisfiable:

$$h = (a \lor b) \land (a \lor \neg b) \land \neg a$$

A huge number of real world problems reduce to SAT, making it possibly the single most important computational problem. As a result, in spite of the hardness of SAT, many decades of research has been devoted to finding efficient algorithms for solving instances of SAT – algorithms that work very fast for *most* instances of SAT. One of the outcomes of this research has been the availability of highly efficient SAT solvers – ones that can handle SAT instances having millions of clauses in less than a second! SAT-based FPV tools harness this efficiency by translating the symbolic reachability problem into an instance of SAT.

3.6.2 SAT-based Reachability

Our strategy for property verification remains the same when we use SAT-based techniques. Given a temporal property, φ, we will search for a counter-example run in the state machine of the implementation, that is, we will look for a run that satisfies $\neg \varphi$.

Let $S = \{s_0, \ldots, s_k\}$ denote the set of state bits of the implementation state machine. These state bits can take different values in different cycles. Let s_j^i be a Boolean variable representing the value of state bit s_j at the ith cycle. For example, s_3^0 represents the initial value of state bit s_3, and s_3^2 represents the value of the same bit after two cycles. Let $S^i = \{s_0^i, \ldots, s_k^i\}$.

We will translate the FPV problem of checking the property, φ, on the given implementation, $\mathcal{M}$, into an instance of SAT over the variables in $\bigcup_i S^i$.

Let us start with an example, namely our priority arbiter of Example 3.1. We shall demonstrate the checking of the property, P1:

$$P1: \quad G[\, r_1 \Rightarrow Xg_1 \land XXg_1 \,]$$

As before, we will look for a run satisfying $\neg P1$ in the implementation. Let $\varphi = \neg P1$, that is:

$$\varphi = F[\, r_1 \land (\neg Xg_1 \lor \neg XXg_1)\,]$$

Our goal is to check whether we are able to reach a state where r_1 is true and g_1 is false in at least one of the next two states. Therefore the shortest witness to this property has two cycles.

We are also given the Boolean functions for the state transition relation of the arbiter implementation:

$$\text{C1:} \qquad (r_2 \wedge \neg r_1 \wedge \neg g_1) \Rightarrow g_2'$$
$$\text{C2:} \qquad r_1 \Rightarrow g_1'$$

The state bits are r_1, r_2, g_1, g_2. Let $S^i = \{r_1^i, r_2^i, g_1^i, g_2^i\}$.

Finally, we are given that initially $g_1 = 0$ and $g_2 = 1$. Therefore the initial set of states is given by the clause:

$$I: \qquad g_2^0 \wedge \neg g_1^0$$

We can use the definition of the transition relation to characterize the set of next states for the initial set of states as follows:

$$C_1^1: \qquad (r_2^0 \wedge \neg r_1^0 \wedge \neg g_1^0) \Rightarrow g_2^1$$
$$C_2^1: \qquad r_1^0 \Rightarrow g_1^1$$

The clause, C_1^1, says that g_2 is true in cycle-1 if r_2 is true and both r_1 and g_1 are false in cycle-0. The clause, C_2^1, says that g_1 is true in cycle-1 if r_1 is true in cycle-0.

Is there any witness of length two for the property φ in the arbiter implementation? To answer this question we need to test whether φ can be satisfied within two cycles under the constraints imposed by $I \wedge C_1^1 \wedge C_2^1$. Witnesses of length two for φ can be characterized by the following clause:

$$Z^1: \qquad r_1^0 \wedge \neg g_1^1$$

The clause, Z^1, says that r_1 is true in cycle-0 and g_1 is false in cycle-1. To test whether such a witness exists in the implementation, we check the satisfiability of the Boolean formula:

$$I \wedge C_1^1 \wedge C_2^1 \wedge Z^1$$

In this case the formula is unsatisfiable, since Z^1 conflicts with C_2^1.

In the next step we ask – *Is there any witness of length three for φ in the arbiter implementation?* Again we use the transition relation to generate the following clauses:

$$C_1^2: \quad (r_2^1 \wedge \neg r_1^1 \wedge \neg g_1^1) \Rightarrow g_2^2$$
$$C_2^2: \quad r_1^1 \Rightarrow g_1^2$$

Witnesses of length at most three for φ can be characterized by the following formula:

$$Z^2: \quad (r_1^0 \wedge (\neg g_1^1 \vee \neg g_1^2)) \vee (r_1^1 \wedge \neg g_1^2)$$

Z^2 describes all ways in which φ can be satisfied in three cycles. The first term specifies all ways in which φ can be satisfied when r_1 is true in cycle-0. The second term is similar to Z^1 for request r_1 arriving in cycle-1. If any of these witnesses are present in the implementation, then the following formula will be satisfiable:

$$I \wedge C_1^1 \wedge C_2^1 \wedge C_1^2 \wedge C_2^2 \wedge Z^2$$

The above formula is indeed satisfiable! The SAT solver will return the following witness (x denotes *"don't care"*):

$$r_1^0 = 1, \; r_2^0 = x, \; g_1^0 = x, \; g_2^0 = x,$$
$$r_1^1 = 0, \; r_2^1 = x, \; g_1^1 = 1, \; g_2^1 = x,$$
$$r_1^2 = x, \; r_2^2 = x, \; g_1^2 = 0, \; g_2^2 = x$$

The witness is actually a set of counter-example traces, where r_1 is high in cycle-0 and low in cycle-1, and g_1 is low in cycle-2. This is indeed a bug in our arbiter implementation, since the property $P1$ requires that g_1 is asserted for *two* cycles following the arrival of the request, r_1.

3.6.3 Detecting Loops

For some types of properties, counter-examples are given by loops in the state machine of the implementation. For example, suppose we wish to verify the property, $f = Fg_1$, which says that g_1 will be eventually asserted. In order to verify the property we need to check whether the implementation has a run that satisfies $\neg f = G \neg g_1$, that is, g_1 is low in every state of the run.

A run of a finite state machine can satisfy $G \neg g_1$ only if there is a cycle in the state machine where every state satisfies $\neg g_1$. Therefore in order to find such a run, we need to detect loops in the state machine.

A run has a loop if the valuation of the state variables, S^i, at some cycle, i, matches with the valuation of the state variables, S^j, at some cycle $j < i$. Therefore we detect a loop within the ith iteration if the following clause is true:

$$loop_i : \quad \bigvee_{j=0...i-1} \left(s_0^i = s_0^j \wedge \ldots \wedge s_k^i = s_k^j \right)$$

3.6.4 Unfolding Properties

The formal rules for generating the clauses that characterize the bounded length witnesses of a property are presented in this section. Suppose the bound on the length of the witness is m. Let $[f]_{j,m}$ denote the set of clauses that must be considered in order to determine whether the property f is true at the jth cycle where $j \leq m$. The following rules recursively define the clauses required for LTL properties.

- $[X\ f]_{j,m}\quad =\quad (j < m)\ \wedge\ [f]_{j+1,m}$
- $[F\ g]_{j,m}\quad =\quad \bigvee_{i=j...m}\ [g]_{i,m}$
- $[G\ f]_{j,m}\quad =\quad loop_m\ \wedge\ \bigwedge_{i=j...m}\ [f]_{i,m}$
- $[f\ U\ g]_{j,m}\quad =\quad \bigvee_{i=j...m}\left([g]_{i,m}\ \wedge\ \bigwedge_{x=j...i-1}[f]_{x,m}\right)$

The use of the loop constraint, $loop_m$ (as defined in the last subsection), for $[G\ f]_{j,m}$ may be noted.

3.6.5 Bounded Model Checking

The SAT-based FPV methodology is iterative in nature – in the ith iteration we look for witnesses of length i, and we reach the ith iteration only if all previous iterations failed to provide a witness (of smaller length). The main steps are:

1. Creating a clause, I, to characterize the set of initial states.

2. Unfolding the state machine of the implementation over time and generating a new set of clauses in each iteration. At the jth iteration, the set of clauses from the implementation are:

$$C^j = \bigwedge_{i=0...j-1} R(S^i,\ S^{i+1})$$

where $R(S^i,\ S^{i+1})$ denotes the clauses relating the variables in S^i with those in S^{i+1} through the transition relation.

3. Unfolding the property to generate clauses that characterize all witnesses of a given length. At the jth iteration, we generate a set of clauses, Z^j, that characterize all witnesses of length less than or equal to j.

4. Use a SAT solver to test whether the clauses generated from the property are satisfiable with the clauses generated from the implementation. At the jth iteration, we test the satisfiability of $I \wedge C^j \wedge Z^j$. If the SAT solver returns success, then we report the witness as the bug in the implementation, otherwise we proceed to the next iteration.

If there are indeed no counter-examples (that is, our implementation is correct) then when do we stop?

The SAT-based approach is iterative in nature – we progressively increase the upper-bound on the length of the witness that we seek in the implementation. In practice, more often than not, the properties are bounded temporal properties, that is, the designer expects the property to be satisfied within a known number of cycles. Therefore it suffices to check whether there are counter-examples of the specified length in the implementation. We can terminate with success if we do not find any counter-example within the given number of iterations.

Most modern FPV tools support a SAT-based *bounded* model checker. In these tools, the user can choose an upper-bound on the number of iterations for the SAT-based checker.

What is the implication of choosing a large bound? The number of clauses given to the SAT solver grows with each iteration – new clauses are added, both by unfolding the implementation and by unfolding the property. The number of clauses added in each iteration can be quite large depending on the size of the RTL. As a result, even the best SAT solver runs into capacity problems when the number of iterations are large for a large RTL.

How does this compare with the BDD-based tools? There are two very notable arguments behind the popularity of SAT-based tools. These are:

1. The majority of the formal properties that arise in practice have a small *depth*, that is, in most cases we expect the property to be satisfied or refuted within a small number of cycles from the point where we start the match.

2. SAT-based approaches can handle designs that are orders of magnitude larger than those that can be handled by BDD-based tools, provided that the bound on the number of iterations is small.

Not surprisingly most commercial FPV tools have a bounded SAT-based tool, and also a generic BDD-based tool.

3.7 Concluding Remarks

In this chapter we have studied the basic methodologies for formal property verification – the way in which an FPV tool works. A commercial FPV tool will have many additional features, and many more algorithms for optimizing the effort of the model checker.

A very important component of any FPV tool is the front-end which reads the implementation in a high-level language (such as Verilog) and extracts the state machine from it. In some of our in-house model checking tools we used a commercial Verilog Design Analyzer which parses the Verilog and creates an object model. Translating the object model into the state machine representation (either BDD or SAT) was one of the most challenging aspects in building the tools. In-house tools used in some industries use a front-end for translating Verilog to state machine formats such as BLIF, and then use the state machine representation in both the BDD-based and SAT-based flows.

There are several other notable methods for FPV, which are beyond the scope of this Chapter. These include:

1. *On-the-fly model checking.* In this approach we compute the product of the state machine with the alternating automata in a depth-first manner. The space requirement of this approach is relatively low, since the depth-first approach ensures that the memory requirement is proportional to the depth of the search. The main problem here is that for large state machines the time required for exhaustive depth-first search is very large. Some recent FPV tools use a combination of on-the-fly model checking and BDD or SAT-based model checking.

2. *ATPG-based FPV.* Automatic Test Pattern Generation (ATPG) techniques have been developed for decades now, and some of the ATPG algorithms are able to handle circuits of large size. ATPG-based FPV techniques attempt to use ATPG to create a set of tests that are relevant for a given property.

   Intuitively the approach is as follows. We translate the checker automaton into an RTL state machine. We then add a circuit which compares the output of the checker automaton with the output of the design-under-test. This comparator produces a 1 whenever the output of the design-under-test matches with the (golden) output from the checker automaton. ATPG finds out all tests corresponding to a stuck-at-0 fault at the output of the comparator.

We have also left out techniques for *word level model checking.* These techniques are useful for verifying properties that contain bit-vectors. There are broadly two schools of research on this topic. One approach is to scale down the bit-vectors to single vectors. For example, most properties of a 64-bit Bus can be scaled down to similar properties for a 4-bit Bus – and it suffices to verify the scaled down version to guarantee that the implementation satisfies the 64-bit version as well. However, bit-scaling must be used with caution and should ideally be supported by an inductive proof.

The other approach is to convert bit-vectors to infinite precision words and then use integer arithmetic to prove the properties. Techniques such as Integer Linear Programming (ILP) has also been used to verify word level data-path properties.

The goal of this chapter was to provide a general awareness of the issues governing formal property verification and the main techniques used in existing FPV tools. While this awareness is expected to equip the validation engineer with a better understanding of the working and the limitations of an FPV tool, it is not sufficient to develop a work plan to get around these limitations. For this purpose new validation aids are needed – tools that will enable the validation engineer to evaluate the correctness and completeness of the specification, tools that will enable her to decompose large properties into smaller ones, tools that will enable her to analyze the feasibility of a formal verification test plan. These issues are in the main focus of this book.

3.8 Bibliographic Notes

In [98], Sistla and Clarke analyzed the complexity of model checking several temporal logics. In this work, they showed that LTL model checking is PSPACE-complete. However, Lichtenstein and Pnueli presented a model checking algorithm for LTL that was exponential in the length of the property, but linear in size of the state machine model [79]. This showed that for LTL properties of limited length, LTL model checking is of linear complexity – which means that model checking is feasible whenever we are able to accommodate the state machine model.

Algorithms for CTL model checking on an explicit state machine model were first presented in [35]. In this work it was shown that CTL model checking works in time linear in the length of the property and the size of the state machine model. It was also shown that for the logic CTL*, which subsumes CTL and LTL, the model checking problem is PSPACE-complete.

The notion of representing temporal logic properties by alternating automata and the automata theoretic style of model checking were introduced by Vardi et al. [105, 74]. On-the-fly automata theoretic model checking techniques have been presented in [43, 55]. Kurshan's book [77] explores the automata theoretic approach in details.

The use of BDDs for Boolean function manipulation was introduced by Bryant [21]. The use of BDDs in symbolic model checking [80] was first proposed in [25] and then improved by the use of partitioned transition relations [26, 28]. It was shown [27] that this approach could handle significantly large state spaces as compared to previous model checking methods.

BDD-based symbolic model checking was first proposed for CTL. In [37], Clarke, Grumberg and Hamaguchi presented a symbolic method for translating LTL model checking into CTL model checking, thereby enabling LTL model checking on BDD-based symbolic representations of sequential circuits.

SAT-based bounded model checking (BMC) was first proposed by Biere *et al.* in [17, 18]. The detailed version of this approach appeared in [41]. Recently the problem of SAT-based *unbounded* model checking has also been studied [71, 81].

Several different FPV techniques based on abstractions have been studied. These include cone-of-influence abstractions [38], counter-example guided abstraction refinement [40, 63], and SAT-based abstraction refinement [42, 107].

Many different techniques for minimizing the state space are used in practice. These include partial order reductions [62, 87], bisimulation equivalent reductions [56], and symmetry reductions [36, 54]

Verification of open systems and compositional verification of collections of open systems have been well studied. The problem of model checking in the presence of an adversarial environment is also known as module checking [73]. In general, we do not assume a purely adversarial environment – we restrict the behaviors of the environment by making assumptions. There has been a significant volume of work on *assume-guarantee* reasoning for open systems [1, 30, 64, 66]. This includes some of our own research on syntactic styles for writing properties for open systems [12].

Boppana *et al.* [20] were the first to propose a model checking technique based on sequential ATPG. Further research on this topic can be found in [2, 68, 96].

Is My Specification Consistent?

We make mistakes while writing the RTL. What is the guarantee that we will not make mistakes while writing assertions?

This is one of the main challenges faced by every chip design company that uses static or dynamic assertion verification in its validation flow. In these early stages of adoption of property verification, the problem is more glaring. This is because a chip design company will have tons of designers who understand Verilog RTL, but relatively few who understand an assertion language. Debugging Verilog RTL has been cultivated for decades. Debugging a set of assertions is new and not fully understood by most.

Practitioners of FPV and dynamic assertion-based verification will admit that errors in coding a formal specification are quite common. Expressing the design intent correctly and accurately in terms of formal properties is a real challenge. Incomplete specifications allow bugs to escape detection, while inconsistent specifications lead to the loss of validation productivity, since the error lies in the specification itself.

There is a common misconception that the main cause of errors in a specification is the syntactic terseness of an assertion specification language – simply put, *we make mistakes because the language is complicated.* This is not really the case. Most of the errors appear because the English language specification itself has conflicting statements, and these conflicts get carried into the formal specification as well.

For example, let us consider the original specification of the priority arbiter of Example 1.1 of Chapter 1. The arbiter has two request lines, r_1 and r_2, and two grant lines, g_1 and g_2. The English language specification of the arbiter was:

1. Request line r_1 has higher priority than request line r_2. Whenever r_1 goes high, the grant line g_1 must be asserted for the next two cycles.

2. When none of the request lines are high, the arbiter parks the grant on g_2 in the next cycle.

3. The grant lines, g_1 and g_2, are mutually exclusive.

These statements are conflicting. The first statement requires g_1 to be high at $t+1$ and $t+2$ when r_1 is high at t. Suppose both r_1 and r_2 are low at $t+1$. Then the second statement requires g_2 to be high at $t+2$. The third statement prevents both g_1 and g_2 to be high at $t+2$. We have a conflict.

If we translate each of these statements correctly (as is) into an assertion specification language, we will have an inconsistent specification. The complexity of the assertion language is not an issue here – the problem lies in the logic of the English specification itself.

As the complexity of the design grows, the number of properties required to express the design intent grows, and with it grows the possibility of such inconsistencies. Beyond a point, human debugging of the specification becomes infeasible because many properties may together create a conflict. We need formal methods to check whether the specification is inconsistent, and if so, to find the set of conflicting properties. In this chapter we will present (a) the types of inconsistencies that are common in specifications, and (b) formal methods for checking whether such inconsistencies are present in a given formal specification. The intent of (a) is to caution the validation engineer about the possible forms of inconsistencies that must be guarded against while writing a formal specification, while (b) should be of interest to the EDA engineer engaged in solving such problems.

Before we proceed any further let us clarify the exact intent of this chapter. The causes behind inconsistent specifications may be broadly divided into two categories, namely:

1. *Logical errors.* These are the errors caused by conflicting or ambiguous statements in the original specification. We may classify them as errors in the design intent, and must be addressed by the design architect.

2. *Coding errors.* These are mistakes made by the validation engineer while coding a property in an assertion specification language. This typically happens due to the validation engineer's oversight, or due to an incorrect understanding of the semantics of the assertion specification language.

Our main focus in this chapter is on the first type. Today, errors of the first type can cause a huge problem in the design of large circuits. Inconsistencies in the architectural specification of the design may not be detected until the RTL for the whole design is written. Detection of an architectural inconsistency at that stage may require rewriting major portions of the RTL with catastrophic consequences. If the architectural properties are written formally and the consistency checks performed *a priori*, then such situations can be avoided more often than not.

Common coding errors in specifications will diminish as the validation engineer becomes more familiar with assertion specification languages. If we take a leaf out of software practices, it will be natural to expect that experienced validation engineers will also make mistakes – only these mistakes will be more complex and harder to debug. This is imminent because assertion specification languages have very powerful constructs that widen the expressive power of the language, but also allow the experienced engineer to write completely unreadable (and un-debuggable) properties. The only way to contain such errors is to impose strict coding guidelines for property specifications. We may be reasonable certain that such coding standards will evolve in the near future.

4.1 Satisfiability and Vacuity

It is known that Boolean satisfiability is NP-Complete – we do not have any known polynomial time solution for testing whether a given Boolean formula is satisfiable. In spite of this hardness, there are excellent Boolean SAT solvers that can solve many large problem instances very fast.

Satisfiability of temporal assertions is not the same as Boolean satisfiability. Temporal assertions are satisfied by *runs* that are valuations of the variables over time. A witness for the satisfiability of a Boolean formula at a given cycle is a valuation of the variables in the formula at that cycle. On the other hand, a witness for the satisfiability of a temporal formula may be a *sequence* of valuations of the variables over multiple cycles.

For example, the following Boolean expression (in SVA) is unsatisfiable:

$$(a \mid\mid b) \ \&\& \ (a \mid\mid !b) \ \&\& \ !a$$

because no valuations of a and b can satisfy this formula at a given cycle. On the other hand, the SVA temporal expression:

$$(a \mid\mid b) \ \#\#1 \ (a \mid\mid !b) \ \#\#1 \ !a$$

is satisfiable – any run where a is true in the first two cycles and false in the third cycle acts as a witness.

4.1.1 Writing Unsatisfiable Specifications

If we are not careful while interpreting the English language specification, we may produce an unsatisfiable specification. Let us consider an example. Suppose we have the following properties for the MyBus master of Section 2.4.2.

1. When the master is not in the IDLE or WAIT states, the request line, **req**, should be kept high.

2. The master does not assert the request line, **req**, all the time.

Suppose we code these properties as shown in Fig 4.1. The specification is unsatisfiable. *Why?*

```
'define IDLE 3'b000
'define WAIT 3'b001

property ReqHighDuringTransfer;
    @ (posedge clk)
    (state != 'IDLE || state != 'WAIT) |-> req ;
endproperty

property ReqIsSometimesLow;
    @ (posedge clk)
    ##[0:$] ! req ;
endproperty
```

Fig. 4.1. An Unsatisfiable Specification

Let us consider the antecedent part of the implication in the first property:

(state != 'IDLE || state != 'WAIT)

This expression will always evaluate to true. The state cannot be IDLE and WAIT at the same time – hence in every cycle, it will either be not equal to IDLE, or it will not be equal to WAIT.

Since the antecedent part of the implication is always true, a master can satsify the first property only if it asserts the request, **req**, all the time. This conflicts with the second property which requires **req** to be low sometime.

Is the English specification incorrect? Not really. Our interpretation of the clause – *the master is not in the IDLE or WAIT states* – was incorrect. The correct interpretation would lead us to write the first property as:

```
property ReqHighDuringTransfer;
    @ (posedge clk)
    (state != 'IDLE && state != 'WAIT) |-> req ;
endproperty
```

This property is consistent with our second property, and they together express the design intent.

4.1.2 The Notion of Vacuity

A property is useful if it can distinguish between correct and incorrect runs. An unsatisfiable specification is false on all runs and is therefore not useful. Similarly a specification that is true on all runs is also not useful. Such specifications are called *vacuous*.

For example, let us consider the following property for the MyBus master device – *When the grant signal*, **gnt**, *is high, the master must not be in the IDLE or WAIT states*. Suppose we make a similar mistake as before, and write this property as:

```
property UseBusWhenGranted;
    @ (posedge clk)
    gnt |− > (state != 'IDLE || state != 'WAIT) ;
endproperty
```

We have made the same mistake, except that the erroneous sequence expression is now in the consequent part of the implication. Since the consequent is always true, the property will always be true.

This is dangerous. The implementation may not actually satisfy the design intent – an invalid MyBus master device may spend idle cycles in the IDLE or WAIT states even after receiving the grant. Our property will be *vacuously* true on all runs and therefore will not flag any error. The validation engineer will believe that the implementation is correct.

Unsatisfiability in specifications is typically detected early because it fails on *every* run of the system. Given an unsatisfiable specification, an FPV tool will return a false counter-example trace. When the validation engineer finds that the counter-example is false, she will know that the specification has a problem.

This is not the case with vacuity. Given a vacuous specification, an FPV tool will simply pass all implementations. The validation engineer will get a false sense of confidence assuming that the properties have passed. Therefore vacuity checks on specifications is an important task in FPV. Just as we need to verify whether a property can at all be satisfied, we need to verify whether it can at all be refuted.

4.2 Satisfiability is not Enough

The roots of the model checking techniques used in FPV tools lie in modeling the design and its environment as a closed system and then verifying properties

on the integrated system. In other words, we treat the input bits also as part of
the state, and use non-determinism to express the freedom of the environment
in setting the values of the input variables. All properties expressed in linear
temporal languages, such as LTL and SVA are said to be true if they are
true on all runs of the system. This automatically means that we check the
property under all possible inputs.

Property specification languages such as LTL, CTL, SVA, PSL, therefore
do not distinguish between the input signals and output signals of the module under test. We can write properties freely using an unrestricted mix of
input and output signals. This freedom is also the source of several forms of
inconsistencies in specifications.

The main issue here is – *a property that is consistent when interpreted over
a closed system can be inconsistent when interpreted over all open systems.*
For example, suppose we have the following requirement for an arbiter:

- *Whenever the high priority req,* hreq, *arrives, the grant line,* hgnt, *is given
  for one cycle with highest priority.*

Suppose we interpret the requirement as – *whenever* hreq *arrives, assert* hgnt
in the next cycle and lower it after one cycle. We will then code this property
as:

```
property HighPriorityGrant;
    @ (posedge clk)
    hreq |−> ##1 hgnt ##1 !hgnt ;
endproperty
```

This property is inconsistent! Suppose hreq arrives in two consecutive
cycles, t and $t+1$. We will have a conflict at time $t+2$, because the request at
t will require hgnt to be lowered at $t+2$, and the request at $t+1$ will require
hgnt to be asserted at $t+2$.

What is the error in this specification? We interpreted the property incorrectly. The phrase – hgnt *is given for one cycle* – does not mean that
hgnt needs to be lowered after one cycle. It only means that the arbitration
is performed again after one cycle. Such errors in interpretations are quite
common in practice, and the errors grow with the complexity of the design
intent. Our goal is to develop formal methods for automatically detecting such
inconsistencies in the specification.

Is the property satisfiable? Yes. This is because we have runs that can
satisfy the property – all runs in which hreq does not arrive in consecutive
cycles satisfy the property.

This example demonstrates an important requirement for the consistency of specifications for open systems, that is, the specification must not only be satisfiable, but it must be satisfiable under *all possible input scenarios*. This is obviously a stricter restriction as compared to satisfiability.

Inconsistencies in open system specifications may also involve multiple properties. For example, let us consider the following properties for a two-input arbiter.

1. The high priority request, hreq, is serviced immediately by asserting hgnt in the next cycle.

2. The low priority request, lreq, must be serviced within the next three cycles.

3. The grant lines are mutually exclusive.

This specification is inconsistent. When the high priority request arrives in three consecutive cycles, a low priority request that arrives in the first cycle cannot be served within the next three cycles. Note that all properties participate in the inconsistency.

It is also interesting to note that the inconsistency in the above specification disappears if we assume that hreq never arrives consecutively in three cycles. If the arbiter is actually to be used in an environment where this is indeed the case, then the specification is correct in the presence of the formal *assumption* that hreq never arrives consecutively in three cycles.

Specifying the right assumptions about the environment is a very significant requirement for performing meaningful formal property verification. In the absense of formal assumptions, the FPV tool may produce false counter-examples covering scenarios that will never occur in practice due to the constraints on the environment.

A consistency check on a specification can reveal inconsistencies in the specification. It may also reveal the need to add *assume* constraints.

4.2.1 Realizability

The formal specification for a module is said to be *realizable* or *synthesizable* if and only if, there exists some implementation of the module that satisfies the specification.

Obviously if a specification is unsatisfiable then it is also unrealizable. In the previous section we also noted that realizable specifications must necessarily be satisfiable under *every given input scenario*. In other words, if we are given an input sequence, then we must be able to provide a valuation of the non-input signals in each cycle, such that the resulting run satisfies the specification.

There is one more intricate issue involved in the realizability of a formal specification. We will explain this through the following example:

```
property Gnt;
    @ (posedge clk)
    req |− > ##[1:$] gnt ;
endproperty
property LowReqAfterGnt;
    @ (posedge clk)
    gnt |− > ##1 !req ;
endproperty
```

The first property says that the request, req, is eventually granted by asserting the gnt signal. The second property says that the req line must be lowered in the cycle following the receipt of the grant.

If we check the specification at the system level with the arbiter and the requesting master device taken together, then we do not have any problem.

Suppose we check the specification only on the arbiter. The req signal is an input to the arbiter and the gnt signal is an output of the arbiter. The first property can be satisfied by any implementation of the arbiter which guarantees that every req is eventually serviced.

The problem lies in the second property. This property should not be a part of the arbiter specification – it actually represents a requirement of the requesting device. *What happens if the validation engineer inadvertently makes it a part of the arbiter specification? Can we detect the inconsistency?*

Can the arbiter satisfy the second property? The answer is yes, but in a peculiar way. Since the arbiter *cannot see its future inputs*, the only way in which it can still satisfy this property is by never asserting the gnt line. This will satisfy the second property, but we will have a conflict with the first property whenever the input line, req, goes high. Therefore if the validation engineer makes the mistake of keeping both properties in the arbiter specification, then the specification will be unrealizable.

This example shows that a perfectly realizable property for one module may be unrealizable for another module where the input/output polarity of the signals are different. A typical assertion IP for a complex protocol will have many assertions – some of these assertions are properties of the whole system, while the others relate to specific devices. For example, the PCI Bus protocol has properties for the master, slave and arbiter devices. It also has system level properties. For example, a property such as – *the bus is never idle when there are one or more pending requests* – is a property which requires specific guarantees from the arbiter (which must not delay the grant), the master (which must float the address on time), and the slave (which must be ready for the transfer). To test a subsystem consisting of one or more devices, we need to choose the correct combination of properties, otherwise the specification may become unrealizable.

Unrealizability is hard to debug manually. In a complex unrealizable specification, it is hard to demonstrate all scenarios under which we get a false counter-example.

We will present the formal definition of realizable specifications later in this Chapter. At the moment let us summarize the necessary and sufficient conditions for a specification to be realizable.

- A specification consisting of properties for an open system (module) having inputs, $\mathcal{I}$, and non-inputs, $\mathcal{O}$ is realizable iff there exists some implementation that satisfies the specification under the following restrictions:

  1. The module is unable to foresee its future inputs.

  2. The module satisfies the specification under all input scenarios.

  If *assume* constraints are given, then we read the second line as – the module satisfies the specification under all input scenarios that are consistent with the given *assume* constraints.

We shall show later that the realizability problem can be interpreted as a game between a hypothetical module and its environment, where the goal of the module is to satisfy the properties and the goal of the environment is to choose the inputs in a way that leads to possible refutation. A specification is realizable if and only if we have a winning strategy for the module.

4.2.2 Receptiveness

In dynamic property verification, we sometimes find that an assertion failure and the corresponding cause for failure do not happen in the same cycle. Often on debugging the cause, we find that the actual fault took place several cycles ago, but was not detected at that time because the specification had no assertion covering that specific behavior.

For example, consider a packet-based real time data transfer protocol (such as PCI XP) where for every *send* event, the sender expects an *acknowledgement* event within some k cycles. Now, a packet may be dropped in any intermediate node, it may be dropped for several reasons, and it may actually happen much before the deadline of k cycles. The assertion which says:

$$\text{send } |-> \text{ \#\#[1:k] acknowledge}$$

will fail only after k cycles (when the deadline expires), and we will have to trace the path followed by the packet from the source to destination to determine where it was dropped and for what reason.

We could detect the fault immediately if we wrote assertions that covered all possible ways in which a packet can be dropped and checked these assertions on all intermediate nodes from the source to the destination. In a complex protocol this can be quite a challenging task because the number of ways in which a packet can be dropped may be quite large. Therefore we often choose to write specifications that do not fail immediately when a fault occurs, but may eventually lead to a failure.

These kinds of gaps in the specification sometimes lead to the masking of a fault before it is detected. We present a more detailed example to demonstrate this point.

Example 4.1. Let us consider the following specification of an arbiter, $\mathcal{A}$, having two request lines, r_1 and r_2, and three grant lines, g_1, g_2 and g_d. The arbiter specification consists of the following properties:

1. P1: The request r_1 is granted in the next cycle by asserting g_1.

2. P2: The request r_2 must be granted within the next 3 cycles by asserting g_2.

3. P3: The grant line g_d goes to the default master (which is a bridge to the low performance bus). g_d must be asserted at least once in every 3 cycles.

4. P4: The grant lines are mutually exclusive.

We also have the following assumptions on the input lines:

1. A1: The request lines, r_1 and r_2 are mutually exclusive.

2. A2: The request r_1 never arrives in consecutive cycles.

3. A3: When r_2 arrives, it remains high until g_2 is asserted.

The SVA code for this specification is shown in Fig 4.2.

Let us now consider the scenario shown in Fig 4.3. At time t, the request line r_1 went high while the grant was parked on g_d. At $t+1$, the arbiter asserts g_1 as required by P1, and also, the request r_2 arrives.

```
property HighPriorityGrant;
    @ (posedge clk) r1 |− > ##1 g1 ;
endproperty
AssertP1: assert property (HighPriorityGrant) ;

property DeadlineForR2;
    @ (posedge clk) r2 |− > ( ##1 g2 ) or ( ##2 g2 ) or ( ##3 g2 ) ;
endproperty
AssertP2: assert property (DeadlineForR2) ;

property DefaultGrantPattern;
    @ (posedge clk) ( gd ) or ( ##1 gd ) or ( ##2 gd ) ;
endproperty
AssertP3: assert property (DefaultGrantPattern) ;

property Mutex;
    @ (posedge clk) (!g1 || !g2) && (!g2 || !gd) && (!g1 || !gd) ;
endproperty
AssertP4: assert property (Mutex) ;

property MutexOnR1R2;
    @ (posedge clk) (!r1 || !r2) ;
endproperty
AssumeA1: assume property (MutexOnR1R2) ;

property NoConsecutiveR1 ;
    @ (posedge clk) (!r1) or (##1 !r1) ;
endproperty
AssumeA2: assume property (NoConsecutiveR1) ;

property PersistsR2 ;
    @ (posedge clk) r2 |− > r2 [*1:$] ##1 g2 ;
endproperty
AssumeA3: assume property (PersistsR2) ;
```

Fig. 4.2. An Unreceptive Specification

At $t + 2$, the arbiter appears to have two options – it may assert g_2 to satisfy P2, or it may assert g_d to satisfy P3. None of these choices violate the given properties – at least not at this moment.

The first option is actually incorrect!! If the arbiter asserts g_2 at $t + 2$, then at $t + 3$ it must assert g_d to prevent P3 from being refuted (recall that g_d was low at $t + 1$ and $t + 2$). Now, if the request r_1 arrives at $t + 2$, then (by P1) we will also need g_1 at $t + 3$. This is not possible by virtue of the mutual exclusion between the grant lines (P4).

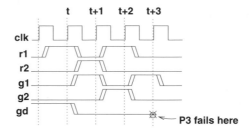

Fig. 4.3. A Conflicting Scenario

On the other hand, if the arbiter asserts g_d at $t + 2$, then r_2 waits (by assumption A3) and thereby r_1 cannot also arrive at $t + 2$ (by assumption A1). Therefore the arbiter will be able to assert g_2 at $t + 3$ without any conflict.

Let us analyze the problem with the first option more carefully. The first interesting thing to note is that the faulty behavior of asserting g_2 instead of g_d at time $t + 2$ will not be exposed by our specification unless r_1 arrives at time $t + 2$. In other words, if r_1 did not arrive at time $t + 2$, the arbiter would have got away with the incorrect response.

Therefore, though the specification does not become *unsatisfiable* in the presence of the fault, it becomes *unrealizable* in the presence of the fault. □

Intuitively, a specification is said to be *receptive* if it becomes unsatisfiable in the presence of every fault for which it becomes unrealizable. In other words, it is never the case that the presence of a fault makes it unrealizable but not unsatisfiable. The specification of Example 4.1 is *unreceptive* because the specification became unrealizable at $t + 2$ under the presence of the fault (asserting g_2 at $t + 2$), but does not become unsatisfiable (it becomes unsatisfiable only when r_1 arrives at $t + 2$).

Can we make the specification of Example 4.1 receptive? The answer is yes. We could add the following property to cover the fault:

```
property WhenToAssertGd;
   @ (posedge clk)
   r1 ##1 r2 |− > ##1 gd ;
endproperty
```

The intent of the above property is hard to fathom unless one is presented with the scenario of Example 4.1. Typically it is quite unrealistic to expect that the validation engineer will identify all such scenarios and add properties to make the specification receptive. Hence we must accept the possibility of dealing with unreceptive specifications.

Unreceptive specifications do not pose any major problem for an FPV tool. However formal methods for testing the receptiveness of specifications are gaining in significance for at least the following two reasons:

1. A specification may become unreceptive if we attempt to verify a property in the wrong context. For example, consider the following property over a request line, req, and a grant line, gnt.

   ```
   property LowReqAfterGnt;
      @ (posedge clk)
      gnt |− > ##1 !req ;
   endproperty
   ```

   The property says that the request line must be lowered in the cycle after receving the grant. This property is not meant for the arbiter, but what if we inadvertently treat this as a property for the arbiter? Interestingly the property can still be realized, but in a rather peculiar way. Since the arbiter cannot see its future inputs, it can satisfy this property by never asserting the gnt signal.

   Is the property receptive? No. If the arbiter asserts the gnt signal, the consequent property, ##1 !req, is unrealizable but not unsatisfiable.

   Inadvertent use of unreceptive properties such as this may not make the specification inconsistent (say, unsatisfiable or unrealizable), but they may not reflect the actual design intent. For example, an arbiter which never asserts a given grant line was not our intent. A receptiveness check can point out the possibility of such mistakes in the specification.

2. In dynamic property verification, a faulty behavior may get masked when the test bench fails to drive the right sequence of inputs that propagates the fault until the unreceptive assertion fails. If we use a realizability checker in the test generation algorithm, then we can intelligently drive the test bench to a refutation whenever the module produces a faulty behavior. The details of this approach is presented in Chapter 7.

Satisfiability, vacuity, realizability and receptiveness are the main consistency issues in formal property specifications. There are other forms of inconsistencies that may be present in a specification, but those are beyond the scope of this book.

4.3 Games with the Environment

The first step towards developing methods for checking satisfiability, realizability and receptiveness of specifications is to formalize these problems. There are many ways to formalize these problems, but we will choose a game theoretic formulation because it demonstrates the differences between these problems very lucidly.

In all of these problems, our target is to verify whether the specification itself has a problem. If so, then there must be some scenario that exposes the problem – our goal is to search for such scenarios. We will model this search as a two player game, where Player-1 represents the module and Player-2 represents its environment. In each round of the game, Player-1 decides the values of the non-input signals, following which Player-2 decides the values of the input signals. In the first round, the values set by Player-1 are the initial values of non-inputs, while the values set by Player-2 are the initial inputs. The module (Player-1) reacts to the inputs received in a given round by choosing its next state, which is manifested by the values of the non-input signals in the next round. The objective of Player-1 is to trace a run that satisfies the specification, while the objective of Player-2 is to guide the run into a refutation.

It is important to note that in this analysis, the module implementation is not given. If Player-1 has a way to win the game, then the synthesis of that winning strategy will be a valid implementation of the module.

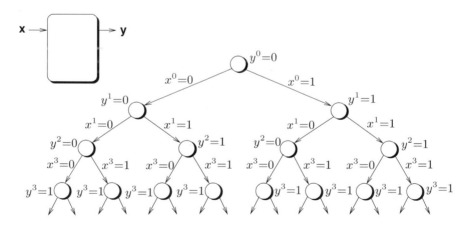

Fig. 4.4. A full-x tree

The *strategy* of Player-1 can be defined over a *full-x tree*, which is a tree of infinite depth that represents all input sequences. For example, a full-x tree for a module having an input, x, and an output, y, is shown in Fig 4.4. The root node represents the initial state of the system. Every other node of the tree represents a state reached by a distinct sequence of inputs from the initial state. The moves of Player-1 can be shown by the values of the non-input signals at each node of the *full-x tree*. These labels define the *strategy* of Player-1. We use y^k and x^k to denote the value of the output y and the input x respectively in the kth round of the game.

Each path of the full-x tree when annotated with the node labels, describe a run. Player-1 attempts to set the node labels in such a way that the resulting run satisfies the specification.

For example, consider the property: *whenever x arrives, y must be asserted in the next cycle*. Fig 4.5 shows an incorrect *strategy* for Player-1 on a full-x tree. In this strategy, Player-1 does not assert y in state s, causing a refutation.

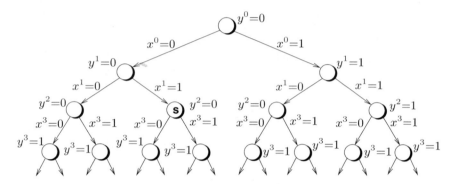

Fig. 4.5. An Incorrect Strategy

Our task of verifying the consistency of a specification is essentially a task of finding out whether there exists a correct *strategy* for Player-1. If not then there can be no valid implementation and hence the specification is itself inconsistent.

In the next section we will use the full-x tree representation to formalize and solve the problems of checking the satisfiability, realizability and receptiveness of specifications.

4.4 Methods for Consistency Checking

In recent times, SAT-based FPV approaches are becoming increasingly popular because of their ability to scale to circuits of decent size. This ability is largely due to recent advances in SAT algorithms and the engineering of the

SAT solver tools. Similar advances are also taking place in checking the truth of Quantified Boolean Formulas (QBF), alternatively called as QBF-SAT.

It has been established that the inherent complexity of realizability and receptiveness checks is harder than that of Boolean satisfiability (SAT). We will show that these problems can be formulated as QBF-SAT problems through some transformations[1]. There are at least two distinct advantages in following this approach, namely:

1. The formulation is very simple and not hard to implement.

2. We automatically benefit from future advances in QBF-SAT solvers.

Our QBF-SAT formulation comes from the full-x tree representation of the game between a hypothetical module and its environment. We will now outline the solutions to each of the three consistency problems – satisfiability, realizability, and receptiveness. We will present the solution methodology around LTL – extensions to languages such as SVA and PSL are nontrivial, but not technically challenging.

The methodology is as follows. In Chapter 3 we had shown that a LTL property, φ, over a set of variables, $\mathcal{AP} = \{x_0, \ldots, x_n\}$, can be *unfolded* over k time steps to create a Boolean formula, $[\varphi]^k$, over $\{x_0^0, \ldots, x_n^0, \ldots, x_0^k, \ldots, x_n^k\}$, where x_j^i represents the value of x_j at the ith time step. Each satisfying valuation of these variables represents a k-length witness of the property φ. In other words, if the Boolean formula, $[\varphi]^k$, is unsatisfiable, then it means that φ can never be satisfied within k time cycles. We studied the use of this kind of unfolding in SAT-based bounded model checking, where the goal was to determine whether the unfolded property and the implementation had a common witness.

On the other hand, for consistency checking, we will verify whether the unfolded property is consistent. The important question here is – *when and how do we terminate?* If we unfold a property up to k steps and find that the resulting Boolean formula is unsatisfiable, we will only know that there are no witnesses of length k or less. *How can we determine whether the property is at all satisfiable, that is whether there exists any k for which the unfolded Boolean formula is satisfiable?*

We will show that there exists an upper bound, k^*, on k. If we do not find any witness within k^* steps, then we will never find a witness. In order to establish that such a bound exists for each of the three consistency problems, we will introduce some formalisms.

---

[1] These transformations were first presented in [94].

4.4.1 Alternating Automata and LTL

In Chapter 3, we studied the notion of *very weak alternating automata* (VWAA). VWAA represent exactly the class of languages that can be defined using LTL formulas. The input to the automaton at each cycle is a valuation of the signals over which the property has been written, and the state of the automaton decides whether the run is accepted or rejected. We used the notion of such a checker automaton while creating the tableau for LTL model checking in Chapter 3. We also studied how this automaton can be *unfolded* over time in SAT-based bounded LTL model checking.

Given an LTL formula $\mathcal{L}$, we construct a VWAA, $A_{\mathcal{L}} = (\Sigma, S, s_0, \rho, F)$. The set of states, S, of the automaton consists of all subformulas of $\mathcal{L}$ and their negation. The input set is $\Sigma = 2^{AP}$ where AP is the set of variables in $\mathcal{L}$ (also called *atomic propositions*). The initial state s_0 represents $\mathcal{L}$ itself.

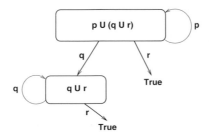

Fig. 4.6. VWAA for $pU(qUr)$ (simplified)

Fig 4.6 presents a graphical representation of the VWAA for the LTL property, $\mathcal{L} = p\ U\ (qUr)$. The nodes represent the subformulas of $\mathcal{L}$. The automata satisfies the following properties by construction:

1. The edges in the graphical representation go from states of higher to lower or same order only, according to the partial order imposed on the states by the relation *"subformula of"*, that is, $n_i \geq n_j$ if n_j is a subformula of n_i . This restriction makes it a *weak alternating automaton (WAA)*.

2. Any cycle in the WAA created from the LTL formula is a self-loop only. This restriction makes it a *very weak alternating automaton (VWAA)*.

The transition function, ρ, of the VWAA maps a state of the VWAA to a subset of S. Consequently a state of the VWAA may be represented by a Boolean formula over the subformulas of the property, $\mathcal{L}$. In other words, the state of the VWAA is represented by a subset of the nodes of its graphical representation, and each transition takes it to a possibly different subset of nodes.

A run of an alternating automaton is a tree rather than a sequence. A finite prefix of this run (say, upto n cycles) may be represented by a propositional formula. This formula can be obtained directly by the following recursive translations, where k is the temporal index (that is, the cycle number).

$$[p]^k = p^k, p \in \Sigma$$
$$[\psi_1 \wedge \psi_2]^k = \psi_1^k \wedge \psi_2^k$$
$$[\neg\psi]^k = \overline{\psi^k}$$
$$[X\psi]^k = \psi^{k+1}$$
$$[\psi_1 U \psi_2]^k = \psi_2^k \vee (\psi_1^k \wedge [\psi_1 U \psi_2]^{k+1})$$

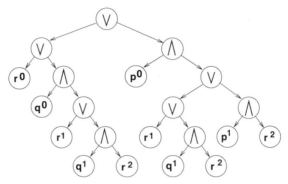

Fig. 4.7. Run tree for $\mathcal{L} = pU(qUr)$ (unfolded to 2 cycles)

Fig 4.7 shows the run tree of the VWAA of Fig 4.6 unfolded upto two cycles. The points to be noted here are as follows:

1. The translations given above create a run tree, representing all possible runs of the alternating automaton of a given LTL formula.

2. The leaves of the tree consist of propositional variables from AP. The internal nodes are labeled by either $\wedge$ (AND) or $\vee$ (OR). The AND-nodes represent universal choices and OR-nodes represent existential choices in the run of the alternating automata.

3. An accepting run is one in which each of the indexed propositional variables have been assigned values that make the propositional formula representing the run tree evaluate to True.

The propositional formula representing the run tree of Fig 4.7 is:

$$\mathcal{F}_{\mathcal{L}}^2 = r^0 \vee (q^0 \wedge (r^1 \vee (q^1 \wedge r^2)) \vee (p^0 \wedge (r^1 \vee (q^1 \wedge r^2) \vee (p^1 \wedge r^2))))$$

Let us now consider an unsatisfiable formula:

$$\mathcal{L} = Gq \wedge F \neg q$$

The propositional formulas obtained by unfolding $\mathcal{L}$ up to zero, one, and two steps are as follows:

$$\mathcal{F}_{\mathcal{L}}^0 = q^0 \wedge \neg q^0$$
$$\mathcal{F}_{\mathcal{L}}^1 = (q^0 \wedge q^1) \wedge (\neg q^0 \vee \neg q^1)$$
$$\mathcal{F}_{\mathcal{L}}^2 = (q^0 \wedge q^1 \wedge q^2) \wedge (\neg q^0 \vee \neg q^1 \vee \neg q^2)$$

None of these formulas are satisfiable, hence $\mathcal{L}$ has no witness of length less than $k = 3$. *Can we conclude that $\mathcal{L}$ is unsatisfiable?* The answer is yes, and the argument is as follows.

Each state s_i of the VWAA of a property $\mathcal{L}$, can be represented by a property $\mathcal{P}_i$, which is a Boolean formula over the subformulas of $\mathcal{L}$. A transition from a state s_i to a state s_j of the VWAA is a *stuttering transition* if $\mathcal{P}_i \equiv \mathcal{P}_j$. A *non-stuttering path* in the VWAA is a sequence of non-stuttering state transitions.

The *"very weak"* property of a VWAA guarantees that a non-stuttering path is a loopless path. It also follows that the length of the longest loopless path is upperbounded by the number of non-self-loop edges in its graphical representation, since (a) each non-stuttering transition of the VWAA must use at least one of these edges, and (b) no edge can be used more than once, since there are no cycles (except the self-loop edges). Therefore the length of the longest loopless path of the VWAA of a property φ is upperbounded by $|\varphi|$, defined as follows:

$$
\begin{aligned}
|\varphi| \quad &= 1 && \text{if } \varphi \text{ is \textbf{true}, \textbf{false}, } p \text{ or } \neg p \text{ where } p \in \Sigma \\
&= 1 + |\psi| && \text{if } \varphi = X\psi \\
&= 1 + |\psi_1| + |\psi_2| && \text{if } \varphi = \psi_1 o \psi_2, \, o \in \{\vee, \wedge, U\}
\end{aligned}
$$

We shall show that unfolding the given property φ up to $|\varphi|$ steps is sufficient to check whether the property is satisfiable, realizable, and receptive.

4.4.2 Satisfiability Checks

An LTL property, $\mathcal{L}$, is satisfiable if there is any run that satisfies the property. The following theorem presents a necessary and sufficient condition for the satisfiability of $\mathcal{L}$ in terms of the propositional formula obtained by unfolding $\mathcal{L}$ to at least the length $|\mathcal{L}|$.

Theorem 4.2. *Let $\mathcal{L}$ be an LTL formula, and let $\mathcal{F}_{\mathcal{L}}$ be the propositional formula representing the run tree of the VWAA of $\mathcal{L}$ that has been unfolded to at least the length $|\mathcal{L}|$. Then, $\mathcal{L}$ is satisfiable iff $\mathcal{F}_{\mathcal{L}}$ is satisfiable.*

Proof: We will prove the stronger result where $\mathcal{F_L}$ represents the propositional formula for the run tree of the VWAA that has been unfolded to at least the length of the longest loopless path in the VWAA.

Only if If $\mathcal{F_L}$ is satisfiable, then there exists a possible assignment of values for the variables at every level of the run tree that make the run tree evaluate to **true**; – that is, an accepting run of the VWAA of $\mathcal{L}$ is possible, therefore $\mathcal{L}$ is satisfiable.

If To prove this we have to show that if $\mathcal{L}$ is satisfiable, that is, if there exists any input sequence $a_1, \ldots, a_n$ to the VWAA that makes $\mathcal{L}$ **true**, then $\mathcal{F_L}$ is satisfiable.

Case1: $n \leq |\mathcal{L}|$ The sequence $a_1, \ldots, a_n$ is an assignment for the variables at level $1 \ldots n$ of the run tree that gives an accepting run, because $a_1, \ldots, a_n$ make $\mathcal{L}$ **true**. This means that this assignment makes the run tree (that is, $\mathcal{F_L}$) evaluate to true. Therefore, $\mathcal{F_L}$ is satisfiable.

Case2: $n > |\mathcal{L}|$ If the length of the accepting run $a_1, \ldots, a_n$ is longer than the length of the longest loopless path in the run tree, it implies that during this run the VWAA *stutters* on certain states. If the sections of the input sequence that represents such *stutters* are removed, we are left with an accepting sequence that is of length no greater than the length of the longest loopless path. So, by the same argument as in *Case1*, this sequence now represents a satisfiable assignment for $\mathcal{F_L}$. Therefore, $\mathcal{F_L}$ is satisfiable. □

To see how this procedure works, let us return to the unsatisfiable property: $\mathcal{L} = Gq \wedge (F\neg q)$. The run tree up to depth 3 is shown in Fig 4.8, and the corresponding propositional formula is:

$$\mathcal{F}_\mathcal{L}^2 = (q^0 \wedge q^1 \wedge q^2) \wedge (\neg q^0 \vee \neg q^1 \vee \neg q^2)$$

The above formula is unsatisfiable, and therefore we conclude that $\mathcal{L}$ is unsatisfiable. Theorem 4.2 guarantees that whenever we have an unsatisfiable prefix of the run tree, whose length exceeds the length of $\mathcal{L}$, then $\mathcal{L}$ is unsatisfiable, and we need not unfold the property any further.

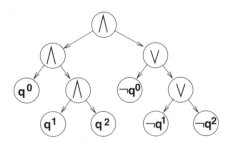

Fig. 4.8. Run tree for $Gq \wedge (F\neg q)$ (unfolded to 3 cycles)

Most FPV tools have the capability to verify whether the specification is satisfiable, though this feature may not be advertised explicitly. Tools which use an approach similar to the tableau based one, can easily determine whether the deterministic checker automaton is itself empty.

The approach presented here is SAT based, which scales well to specifications of large size. Also, unlike in the case of the model checking problem where we compare the specification with an implementation, the satisfiability problem does not have to deal with any implementation and is therefore largely free from the state explosion problem. Capacity issues may however come up if we use large bit vectors in the specification, or if our properties have large temporal depth (such as timed properties with large time bounds). We tested the methodology on the ARM AMBA Bus protocol suite after scaling down the Bus width, and did not face any capacity problems.

4.4.3 Realizability Checks

In order to verify whether a given specification is realizable we need to know the direction (input / ouput) of the signals. In Section 4.3 we studied the notion of a full-x tree in the context of viewing the task of consistency checking as a game between a hypothetical module (Player-1) and its environment (Player-2). In that model, Player-1 sets the values of non-inputs in every round, while Player-2 sets the values of the inputs. Player-1 reacts to the move of Player-2 by choosing its next state, which is manifested by the values of the non-inputs in the next round. Intuitively, a round of the game corresponds to a clock cycle of the module, where the present state (that is, present values of state-bits) and present inputs decide the value of the next state of the module as per its implementation strategy.

Let us first see how satisfiability checking fits into this picture. A specification is satisfiable if there exists at least one run that satisfies the specification. This is equivalent to saying that at least one path of the full-x tree is a valid run.

Let us consider our earlier example of an unrealizable (but satisfiable) property (now written in LTL) where r is a high priority request and g is the corresponding grant:

$$\mathcal{L} = G[\ r \ \Rightarrow\ Xg \ \wedge \ XX \ \neg \ g \ ]$$

Fig 4.9 shows one strategy of Player-1 on a full-x tree representation. With this strategy, the leftmost path, $\pi = 1, 2, 4, \ldots$, of the full-x tree is a witness for $\mathcal{L}$.

Therefore, when we want to test the satisfiability of the specification we will assume that the module and its environment cooperate to find some

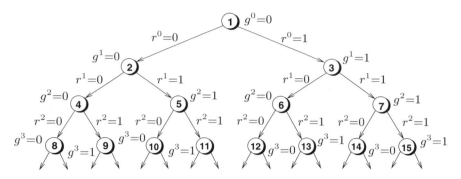

Fig. 4.9. One strategy for G[r ⇒ Xg ∧ XX ¬ g]

valuation of the signals over time, such that the resulting run satisfies the specification. In other words, at the kth iteration we verify the truth of the property:

$$\exists g^0 \, \exists r^0 \, \ldots \, \exists g^{k-1} \, \exists r^{k-1} \, \exists g^k \, \mathcal{F}_\mathcal{L}^k$$

where $\mathcal{F}_\mathcal{L}^k$ is the propositional formula obtained by unfolding $\mathcal{L}$ up to k steps. Since all the variables are existentially quantified, this is essentially a Boolean satisfiability problem.

When we test the realizability of $\mathcal{L}$, we must treat the module and its environment as adversaries, since we need to verify whether there exists any strategy that can always guarantee the satisfaction of $\mathcal{L}$. In other words, with the desired strategy, every path of the full-x tree must be a satisfying run. Therefore at the kth iteration we verify the truth of the QBF:

$$\exists g^0 \, \forall r^0 \, \ldots \, \exists g^{k-1} \, \forall r^{k-1} \, \exists g^k \, \mathcal{F}_\mathcal{L}^k$$

where $\mathcal{F}_\mathcal{L}^k$ is as before. For example, if we unfold $\mathcal{L}$ twice, we obtain:

$$\mathcal{F}_\mathcal{L}^2 = \quad (r^0 \; \Rightarrow \; g^1 \wedge \neg g^2) \; \wedge \; (r^1 \; \Rightarrow \; g^2)$$

The property:

$$\exists g^0 \, \forall r^0 \, \exists g^1 \, \forall r^1 \, \exists g^2 \, \mathcal{F}_\mathcal{L}^2$$

is not valid, since $\mathcal{F}_\mathcal{L}^2$ is unsatisfiable for $r^0 = r^1 = 1$. This corresponds to the rightmost path, $\pi = 1, 3, 7, \ldots$, of the full-x tree of Fig 4.9. Every valuation of g at node 7 is invalid.

Our algorithm for realizability checking for a LTL property $\mathcal{L}$ works as follows.

1. We first construct the propositional formula $\mathcal{F}_\mathcal{L}^k$ equivalent to the run tree up to depth k, where k is the length of the longest loopless path in the VWAA of $\mathcal{L}$.

2. We then create the QBF:

$$\mathcal{Q_L} = \exists o^0 \; \forall i^0 \; \ldots \; \exists o^{k-1} \; \forall i^{k-1} \; \exists o^k \; \mathcal{F}_{\mathcal{L}}^k$$

where i^j denotes the input vector at step j, and o^j denotes the valuation of the non-inputs at step j. The universal quantification $\forall i^j$ represents all possible valuations of the inputs by the environment in the jth cycle (that is, over input variables indexed by j). The existential quantification $\exists o^j$ represents the existence of a legal valuation of the non-inputs by the module in the jth cycle. Since a module cannot see its future inputs, the quantifiers alternate over the different time cycles. We use a QBF solver to solve $\mathcal{Q_L}$.

The key result here is the following theorem, which shows that the property, $\mathcal{L}$ is realizable if and only if the finite length QBF $\mathcal{Q_L}$ is true.

Theorem 4.3. *Let $\mathcal{L}$ be an LTL formula, and let $\mathcal{F}_{\mathcal{L}}^k$ be the propositional formula representing the run tree of the VWAA of $\mathcal{L}$ that has been unfolded to at least the length, k, of the longest loopless path in the VWAA. Then, $\mathcal{L}$ is realizable if and only if the QBF $\mathcal{Q_L} = \exists o^0 \; \forall i^0 \; \ldots \; \exists o^{k-1} \; \forall i^{k-1} \; \exists o^k \; (\mathcal{F}_{\mathcal{L}}^k)$ is* **true**.

Proof: *Only if.* To prove this we need to show that if $\mathcal{Q_L}$ is not **true**, then $\mathcal{L}$ is not realizable.

If $\mathcal{Q_L}$ is not **true**, then for each strategy of Player-1, there exists a valuation of inputs set by Player-2 up to some depth j, $j < k$ for which the formula $\mathcal{F}_{\mathcal{L}}^k$ becomes unsatisfiable. This input sequence represents a path in the full-x tree to a node n at depth j where the strategy of Player-1 fails. Therefore, $\mathcal{L}$ is not realizable.

If. If $\mathcal{L}$ is not realizable then for each strategy of Player-1, there exists a node n in the full-x tree where Player-1 has no way of setting its outputs to make $\mathcal{L}$ **true**. This follows from the definition of realizability. We need to show that such a node always exists at a depth j, $j \leq k$ and that $\mathcal{Q_L}$ is **false**. Let j denote the depth of n.

Case1: $j \leq k$ The input valuation along the path to n will make $\mathcal{Q_L}$ false.

Case2: $j > k$ Since $\mathcal{F}_{\mathcal{L}}^k$ is the unfolding of $\mathcal{L}$ up to the length of the longest loopless path in the VWAA, any run of length greater than k will loop at one or more states of the VWAA. In these stuttering states, if we consider only the input valuation that leads towards the node n, then we have the sequence of input valuations that makes $\mathcal{Q_L}$ false. $\square$

4.4.4 Receptiveness Checks

Intuitively, a module is receptive if it does not need to consider possible future inputs while choosing the values of its outputs at a given instant. In other words, it should be free to choose any valuation of its outputs as long as that valuation does not refute the property at that time instant. We shall refer to such valuations as *legal valuations* of the outputs. The notion of receptiveness can be expressed using the full-x tree model as follows.

Definition 4.4. *A property φ is receptive iff at each node of the full-x tree, every legal valuation of the outputs represents a winning strategy for the module.* $\square$

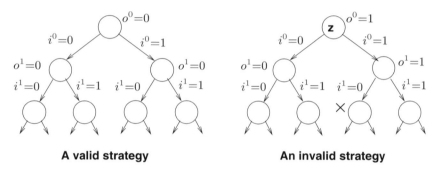

A valid strategy   **An invalid strategy**

Fig. 4.10. Full-x tree for $G(o \Rightarrow Xi)$

Example 4.5. Consider the following non-receptive property for a module with input i and output o:

$$\psi = G(\ o \ \Rightarrow \ Xi\ )$$

Figure 4.10 shows two legal labellings of the node z on the full-x tree of ψ. In the second tree, the module assigns $o = 1$ at z, which is perfectly legal at z, but leads to a refutation if the input i is not asserted in the next cycle. On the other hand, if the module assigns $o = 0$ at all nodes (as in the first tree), then there is no such problem. Therefore the module cannot choose its outputs from the entire set of legal outputs, and we conclude that the specification is not receptive. $\square$

To check for the receptiveness of a property $\mathcal{L}$, we first construct the propositional formula $\mathcal{F}_{\mathcal{L}}^k$ equivalent to the run tree up to depth k, which is the length of the longest loopless path in the VWAA of $\mathcal{L}$. For each j, $j \leq k$, we create a QBF as follows:

$$\mathcal{R}_{\mathcal{L}}^j = \forall o^0 \; \forall i^0 \; \dots \; \forall o^{j-1} \; \forall i^{j-1} \; \forall o^j \; \left(\mathcal{P}_{\mathcal{L}}^j \Rightarrow \mathcal{Q}_{\mathcal{L}}^j \right)$$

where:

$$\mathcal{P}_{\mathcal{L}}^j = \exists i^j \; \exists o^{j+1} \; \dots \exists i^{k-1} \; \exists o^k \; \left(\mathcal{F}_{\mathcal{L}}^k \right)$$
$$\mathcal{Q}_{\mathcal{L}}^j = \forall i^j \; \exists o^{j+1} \; \dots \forall i^{k-1} \; \exists o^k \; \left(\mathcal{F}_{\mathcal{L}}^k \right)$$

The following theorem shows that a property is receptive iff $\mathcal{R}_{\mathcal{L}}^j$ is true for all $j \leq k$.

Theorem 4.6. *A satisfiable property, $\mathcal{L}$ is receptive if and only if for every j, $j \leq k$, $\mathcal{R}_{\mathcal{L}}^j$ is true, where k is the length of the longest loopless path in the VWAA of $\mathcal{L}$.*

Proof: *Only if* To prove this we need to show that if for some instantiation, θ_j, of $o^0, i^0, \dots, i^{j-1}, o^j$, $\mathcal{P}_{\mathcal{L}}^j$ is true and $\mathcal{Q}_{\mathcal{L}}^j$ is false, then $\mathcal{L}$ is not receptive.

Since $\mathcal{P}_{\mathcal{L}}^j$ is true, the valuation of o^j in θ_j is a legal valuation. On the other hand, since $\mathcal{Q}_{\mathcal{L}}^j$ is false, it shows that a module does not have a winning strategy at the node of the full-x tree that is reached by applying θ_j. Therefore, $\mathcal{L}$ is not receptive.

If To prove this we need to show that if $\mathcal{L}$ is not receptive, then there exists a substitution, θ_j, of $o^0, i^0, \dots, i^{j-1}, o^j$ such that $\mathcal{P}_{\mathcal{L}}^j$ is true, but $\mathcal{Q}_{\mathcal{L}}^j$ is false.

Since $\mathcal{L}$ is not receptive, there exists a node n in the full-x tree at some depth j, where for some legal valuation of o^j, there is no winning strategy of the module. We will construct the desired θ_j as follows.

Case1: $j \leq k$ In this case, θ_j is simply the valuation of $o^0, i^0, \dots, i^{j-1}, o^j$ that takes us from the root of the full-x tree to node n. This valuation is a witness for $\mathcal{P}_{\mathcal{L}}^j$. Since there is no winning strategy for the module at node n it follows that $\mathcal{Q}_{\mathcal{L}}^j$ is false.

Case2: $j > k$ Since $\mathcal{F}_{\mathcal{L}}^k$ is the unfolding of $\mathcal{L}$ up to the length of the longest loopless path, any run of length greater than k will loop at one or more states of the VWAA. In these stuttering states, if we consider only the valuations of the inputs that takes us towards node n, we get the desired substitution. The rest of the argument is similar to that of Case1. $\square$

4.5 The SpecChecker Tool

We have implemented the methods outlined in this chapter in a tool that checks the satisfiability, realizability and receptiveness of LTL specifications.

We use *zchaff* [83] as the backend SAT solver and *QuBE* [61] as the backend QBF checker.

The propositional formula representing the longest loopless path in the VWAA for the given LTL specification is first checked for satisfiability using *zchaff*. It is then converted to CNF using the blocking clauses technique described in [81], which is then further tested for realizability using QuBE.

If a property is unrealizable it is not receptive (by definition). Otherwise, we test for receptiveness by checking the truth of at most $|\mathcal{L}|$ QBF formulas.

We also developed a rudimentary prototype tool for checking the realizability of SVA properties. This tool supports only a fragment of SVA – for example, multiple clocks, local variables, etc are not supported. We used a SystemVerilog Design Analyzer to read the SVA properties and the direction of the signals from a SV design, and pass it to the SpecChecker tool.

4.6 Concluding Remarks

We believe that consistency problems in formal specifications will soon become one of the dominating issues in FPV. The task of debugging specifications will require an arsenal of new formal methods.

In this chapter we studied some of the most common forms of logical inconsistencies in formal specifications. Current tools support satisfiability checks (either implicitly or explicitly), but realizability and receptiveness checks are not supported, though these problems are quite common in large specifications. For example, we were pleasantly surprised to find several consistency problems in an industry standard assertion IP for the ARM AMBA AHB protocol – after the assertion IP had passed several rounds of testing, both by its vendor as well as other customers. The problems had remained because the scenarios triggering the realizability and receptiveness problems never came up during (simulation-based) testing. The formal analysis immediately detected the problems.

There is one more important issue that we somewhat glossed over. *How do we demonstrate the inconsistency in the specification?* If the consistency checker simply reports that the specification is inconsistent, and provides no other feedback, then the validation engineer will have to do the debugging manually. *Can the tool also assist the debugging process?*

Let us examine each case separately:

1. *Demonstrating un-satisfiability.* If the specification is unsatisfiable, then we have the advantage that the specification will be refuted on all runs –

including those valid runs which the validation engineer expects the specification to accept. In the debugging process, we may ask the validation engineer to present any run (trace) which she expects the specification to accept. We may then list the properties that fail in that trace and also indicate the specific points (on the trace) at which they fail.

Could we have produced a false counter-example trace directly from the unsatisfiable specification? The answer is no, because we have no way to determine the design intent from the unsatisfiable specification – that is, we cannot distinguish between the valid and invalid runs with respect to the actual design intent. Therefore the sample valid run must come from the validation engineer.

2. *Demonstrating un-realizablity.* This is a much harder problem because an unrealizable specification is *not* refuted on all runs. For example, consider the unrealizable property, $o \Leftrightarrow i$, for a module with input i and output o. If we assert $o = 1$ in the first cycle, then the specification becomes unsatisfiable only when we have $i = 0$ in the next cycle.

Can we produce an input seqeunce under which the specification becomes unsatisfiable? Not always. For example, if we produce an input sequence where $i = 1$ in the second cycle, then the strategy which asserts $o = 1$ in the first cycle will satisfy the specification. On the other hand, if we produce an input sequence where $i = 0$ in the second cycle, then the strategy which asserts $o = 0$ in the first cycle will satisfy the specification. Therefore, the witness for unrealizability is not a run or an input sequence – *it is a strategy of the environment for refuting the specification* – a strategy that always wins, regardless of the implementation strategy.

How can we then demonstrate unrealizability? We can produce an intelligent test bench that plays the winning strategy of the environment. Simulating any implementation with this intelligent test bench will then demonstrate the refutation immediately. We shall study this approach in more details in Chapter 7.

3. *Demonstrating un-receptiveness.* An unreceptive specification is not an inconsistent one – it is merely not tight enough to be refuted in the same cycle as the fault during dynamic verification. Therefore we can demonstrate un-receptiveness when the presence of a fault renders the specification unrealizable, but not unsatisfiable. When this happens, we can use intelligent test generation to drive the simulation towards eventual refutation. This approach serves two purposes – (a) it shows that the specification is unreceptive, and (b) it overcomes the main limitation of unreceptive specifications by not allowing a fault to get masked. We shall elaborate on this approach in Chapter 7.

Early detection of inconsistencies in formal specifications will largely benefit any design validation flow. Appropriate ways of demonstrating the problem in the specification will be an equally important problem. In this Chapter we have outlined only some of the major issues – the subject as a whole is loaded with opportunities for EDA and CAD companies.

4.7 Bibliographic Notes

The problem of determining the realizability of a specification for a sequential circuit was formulated by Church in 1965. In [93], Rabin provided a first solution to Church's problem based on automata over infinite trees. At the same time, Buchi and Landweber provided another solution [23] based on infinite games. Most of the techniques for verifying synthesizability of specifications are built around these two approaches.

In 1989, Pnueli and Rosner [91] formalized the problem of verifying the realizability of specifications for reactive modules, that is, open systems that must satisfy given properties under all possible inputs from the environment, *without being able to foresee the future inputs.*

The receptiveness problem was first formulated by Dill in his thesis [48]. Similar consistency issues in temporal specifications has also been studied by Gawlick *et. al.* [60], and Abadi and Lamport [1].

In 2002, Alur, Henzinger and Kupferman showed that consistency problems for reactive module specifications such as realizability can be expressed in their new logic called Alternating-time Temporal Logic (ATL) [7].

In 2005, we showed [94] that the problems of satisfiability, realizability and receptiveness of temporal specifications can be reduced to *quantified Boolean formulas* (QBF) and solved using efficient QBF solvers.

The notion of *vacuity* in temporal logic model checking has been studied in [75, 76, 11] and is a subject of considerable recent interest.

5

Have I Written Enough Properties?

Logical bugs like to hide in the gap between the design intent specification and the implementation. The RTL designer typically receives the specification as an English document and develops the implementation on the basis of her understanding of this document. Using a natural language such as English creates the possibility of a gap between the design architect's actual intent and the RTL designers' perception of this intent. Some of the hardest logical bugs love to hide in this gap.

The goal of formal property verification is to present the specification in a formal language so that (a) there is no gap between the architect's specification of the design intent and the RTL designer's understanding of that intent, and (b) automatic verification techniques can check whether the implementation meets the specification.

Unfortunately the limitations of existing FPV techniques allow us to achieve this objective only partially. Specifically:

1. It is not practical to express the entire contents of a design specification document in terms of formal properties. One of the major challenges in developing an assertion suite is to identify the right kind of *properties* from the specification document.

2. We cannot formally verify every property that we write because of capacity limitations of existing FPV tools. Therefore, we must attempt to express the design intent through small properties. If this is not possible (as is often the case) then we either resort to dynamic verification (which is non-exhaustive in practice) or we live with an incomplete formal specification.

It is natural to expect that practical considerations will force us to work with formal specifications that cover the design intent partially.

Therefore it is hard to define the meaning of the question – *Have I written enough properties?* Any engineer engaged in developing an assertion suite looks for a suitable answer to this question, and the related question – *What fraction of the design functionality have I covered through my properties?*

In order to answer these questions we need to define the *"whole"* that is at the denominator of this fraction. *What is the "whole" that we aim to cover?* This is the first challenge in formalizing a FPV coverage metric – defining the coverage goal.

There is a popular misconception about FPV coverage. Since a formal property specification models the design intent, many practitioners of FPV expect FPV coverage metrics to indicate *functional coverage*. This expectation creates the following paradox. Our objective is to write a set of properties that expresses the design intent completely. In order to formally ascertain the functional coverage of this property suite, we need the complete design intent (the coverage goal) to be given formally, which in turn is the same as our objective!!

In other words, since the property suite represents the first formal functional specification of a design intent, we do not have any other golden functional model that could act as the reference for functional coverage analysis. It is therefore not surprising that all existing FPV coverage metrics are structural in nature.

A structural coverage metric is not meaningful unless we can relate it to the functional behaviors in some way. Typically, our expectations from structural FPV coverage metrics are as follows:

1. A low structural coverage should always indicate low functional coverage. Therefore a low value of coverage should prompt the designer to write more properties.

2. The coverage report should indicate the coverage gaps in such a way that the validation engineer can easily relate it with the gaps in functional coverage. A coverage result that cannot be related to the functional gaps in the specification is not of much value.

3. A high structural coverage does not guarantee high functional coverage.

There are a couple of notable issues here. Experience from simulation coverage metrics suggest that using more than one coverage metric usually provides a better coverage feedback than using only one. This is natural, since each structural coverage metric approximates the functional coverage goal in a different way. Early experience from practitioners of FPV coverage also seems to indicate that there is value in using multiple FPV coverage metrics to analyze the completeness of a specification. This seems to be the best available option for getting around the limitation indicated in the third point above.

There is also the issue of computational complexity. Many of the existing FPV coverage metrics have the same complexity as model checking. If we have to spend 8 hours in checking a property, and another 8 hours for assessing the completeness of the specification, then the overhead becomes annoying. Moreover, property suites are not developed solely for FPV – assertion IPs for large protocols, such as PCI XP, ARM AMBA Bus, etc, are also developed for dynamic assertion-based verification at the system level. Large system level specifications cannot be used in model checking (due to capacity limitations), but it is very important to assess the completeness of these specifications for the success of dynamic ABV. Therefore we need coverage metrics that can be evaluated quickly, and we need coverage analysis methods that do not depend on a model checking tool.

This chapter outlines a cross section of recent approaches towards formal verification coverage, including some of our recent research on this subject.

5.1 Simulation Coverage Metrics

Simulation coverage metrics have been used for ages. However, the significance of these metrics have grown significantly in recent times due to the adoption of coverage driven randomized test generation. The new approach alleviates the effort of writing directed tests to a large extent and is quite effective in practice, provided that the coverage monitors help in directing the test generation process towards the less covered areas.

Simulation coverage metrics and FPV metrics have a common goal – both aim to cover the interesting behaviors of the design under test. However the method in which these metrics are evaluated are grossly different, and the interpretation of the coverage results are also handled in a different way. It is quite informative to compare the simulation coverage metrics with FPV coverage metrics in terms of their semantics. We summarize some of the most popular simulation coverage metrics in this section.

1. *Code coverage.* The most widely used code coverage metrics are *statement* and *branch* coverage. A statement or a branch of the HDL code is said to be covered if these are executed during the simulation. In constraint-driven randomized test generation we use the feedback to ensure that the enabling condition of each branch of the program is satsified one or more times.

2. *Circuit coverage.* These metrics check the coverage of the parts of the circuit structure during simulation. The most popular forms of circuit cover-

age metrics are *latch* and *toggle* coverage. A latch is covered if it changes value at least once during simulation. Similarly an output is covered if it changes value at least once during simulation.

The advent of *hardware verification languages* has led to the evolution of some new types of simulation coverage metrics. These metrics are becoming popular in the context of coverage-driven randomized test generation and dynamic assertion-based verification.

1. *FSM coverage.* Recent advances in the test bench language enables the validation engineer to specify abstract state machine models of the system and then automatically direct the test generation towards covering all the states and transitions of the FSM. For example, we may create a transaction level state machine model for a Bus protocol where each path represents a distinct type of transfer, and then verify whether all branches of the state machine are covered during simulation.

2. *Assertion coverage.* During dynamic assertion-based verification, we also need to check whether every property has been matched non-vacuously in the simulation run. If not, then some interesting behavior, for which the property was written, did not come up during simulation. Recent property specification languages support automatic evaluation of such coverage metrics – recall the notion of *"cover" properties* in SVA.

3. *Mutation coverage.* This is also a structural coverage metrics, where the validation engineer introduces a fault into the design implementation (HDL code or circuit), and then checks whether this leads to an erroneous behavior. In mutation coverage, the goal is to find a set of tests such that for each mutant design, there exists at least one test that fails in it.

In the next section we will present FPV coverage metrics and show how they compare with the above simulation coverage metrics.

5.2 Mutation-based FPV Coverage

The notion of mutation coverage unifies simulation coverage metrics and FPV coverage metrics. The idea is simple. If some part of the design can be mutated without any side effects on the validity of the design, then either that part of the design is redundant, or our specification (that is, our definition of "validity") is incomplete. Let us begin with an example.

Let us consider a two-way round-robin arbiter, where no requesting device is granted twice in consecutive cycles. Fig 5.1 shows an abstract state machine

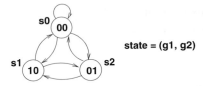

Fig. 5.1. Abstract FSM of round-robin arbiter

extracted from the arbiter implementation. Suppose the specification has the following properties:

1. P1: g_1 *is never asserted in two consecutive cycles.* This may be written in SVA as:

```
property NoConsecutiveG1;
    @ (posedge clk) g1 |-> ##1 ! g1 ;
endproperty
```

2. P2: g_2 *is never asserted in two consecutive cycles.* This may also be similarly expressed in SVA.

Both properties hold on the state machine of Fig 5.1. In order to verify these properties, our search for a counter-example takes us through each state. For example, to verify the first property, P1, we visit each state to check whether it satisfies g_1, and if so, whether it has any next state that satisfies g_1. Since we explore every state, does this mean that we have achieved 100% state coverage?

The answer is obviously negative. Otherwise, every universal property (or invariant) will achieve 100% coverage. The real question should be – *What part of the design contributed to the success of the property?*

If some part of the design contributed to the success of a property, then we may expect that the truth of the property will change if we mutate that part of the design. This is the main idea behind mutation-based FPV coverage metrics.

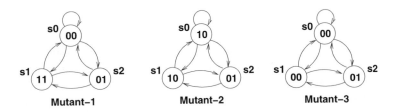

Fig. 5.2. Three mutants of the arbiter FSM

Let us return to our example. Fig 5.2 shows three mutations of the state machine of Fig 5.1. We explain the notion of mutation coverage through these examples.

- In the first mutation, we have toggled the value of g_2 from 0 to 1 in state s_1. As a result, the mutated FSM does not satisfy P2. We therefore conclude that the value of g_2 at s_1 is covered by P2.

- In the second mutation, we have toggled the value of g_1 from 0 to 1 in state s_0. The mutated FSM does not satisfy P1, and we conclude that the value of g_1 at s_0 is covered by P1.

- In the third mutation, we have toggled the value of g_1 from 1 to 0 in state s_1. *This time, all properties remain satisfied!* Therefore, we conclude that the value of g_1 is not covered at state s_1.

The coverage gap discovered through the third mutation can be interpreted as follows – *there is no property that requires g_1 to be asserted.* We can therefore attempt to close the gap by adding a new property that specifies the cases where g_1 needs to be asserted.

How complex is it to evaluate the coverage? It may be noted that the coverage estimation algorithm needs to use FPV techniques to determine the truth of the property in the original FSM and in the mutant FSM. Therefore, if the FPV tool runs into capacity issues, then we cannot assess the coverage of the specification.

This is a serious limitation. Our requirement for analyzing the completeness of a formal specification goes beyond FPV. Dynamic assertion-based verification techniques also use formal property specifications – in fact, this approach is more popular today than FPV. Assessing the completeness of assertion IPs for dynamic and semi-formal property verification techniques is an important requirement which needs to be solved without being hindered by the limitations of existing FPV tools. We will address this requirement later in this chapter.

5.2.1 Falsity and Vacuity Coverage

Once we have a mutant FSM, we can perform two types of coverage checks on it. These are:

1. *Falsity coverage.* In this approach, we check whether the mutant FSM satisfies the specification. If not, then the mutation is covered by the specification. Our earlier examples are instances of falsity coverage checks.

2. *Vacuity coverage.* In this approach, we check whether the mutant FSM satisfies the specification *vacuously*, if at all.

To study the notion of vacuity coverage, let us return to the third mutation of Fig 5.2. *Does this mutation satisfy P1 vacuously?* The answer is, yes. The antecedent part of the implication in P1, namely g_1, does not match any of the states of the mutant FSM. As a result, the property is vacuously satisfied everywhere. In the original FSM, this was not the case since the antecedent, g_1, matched at state s_1. Therefore we conclude that the value of g_1 at s_1 is *vacuity covered* by P1.

If a property matches vacuously at all states, then there are two possibilities, namely:

1. The property itself is vacuous (or valid), and therefore useless for verification.

2. The implementation is incomplete, since it models none of those behaviors for which the property was written.

Vacuity coverage is based on the notion that any mutation that removes the behavior for which a property was written is covered by the property. Obviously if a mutation is falsity covered, then we need not check vacuity coverage on that mutant. On the other hand, if a mutation is not falsity covered, then we may use vacuity coverage to provide more useful feedback.

5.2.2 FSM Coverage

Since FPV works on the state machine model of the implementation, the natural origin of FPV coverage lies in analyzing mutation coverage on the states and transitions of the FSM extracted from the implementation. All the examples presented so far in this chapter are forms of FSM coverage.

Broadly, three different types of FSM coverage metrics are used in practice. These are:

1. *State bit coverage.* This is computed by toggling state bits at different states and checking whether the mutant FSM satisfies the specification. We may use both falsity coverage and vacuity coverage.

2. *Transition coverage.* We remove a transition from the FSM and check whether the mutant FSM satisfies the specification. Universal properties (such as all LTL properties, or all SVA properties) are not affected by the removal of transitions from the FSM, hence falsity coverage is useless here. We use vacuity coverage to determine the coverage of transitions.

3. *Path coverage.* In path coverage we verify the effect of removing or mutating a finite path on the satisfaction of a property.

It is possible to imagine several variants of these metrics, but we leave it to the imagination of the reader!

5.2.3 Code and Circuit Coverage

Recent research has also led to FPV coverage metrics that are semantically similar to the simulation coverage metrics, such as code and circuit coverage. In these approaches, mutations are performed on the HDL code or the circuit structure, and we check whether the truth of the specification changes in the mutant implementation. We briefly explain these methods:

1. *Code coverage.* We extract the control flow graph (CFG) from the HDL code of the implementation. We say that a statement of the CFG is covered if the truth of the specification in the mutant changes when we omit the statement. Removing a branch of the CFG does not affect universal properties (such as LTL or SVA properties), hence we use vacuity coverage for checking the coverage of branches of the CFG.

2. *Circuit coverage.* In simulation-based circuit coverage metrics we verify whether every signal and latch toggles during the execution of the HDL. In the corresponding FPV coverage metrics, we fix the value of a signal (or a latch) and verify whether the truth of the specification changes in the mutant where we do not allow that signal (or latch) to change value.

These metrics have evolved mainly because validation engineers who are familiar with code and circuit coverage metrics, sometimes find it hard to relate to FSM coverage metrics and interpret the result in terms of functional coverage.

It should always be kept in mind that the goal of FPV coverage metrics is to verify whether the *specification*, and not the *implementation*, is complete. Verifying whether the specification covers the implementation is only a means for finding out whether more properties need to be written. Therefore the gaps identified by the coverage metrics should reflect the types of behaviors for which properties should be written. This goal is often not easy if we provide coverage feedback in terms of code or circuit coverage. Our attempt to develop coverage metrics that are close to simulation coverage metrics should not take us far from our main objective of providing useful feedback to the architect of an assertion IP.

5.3 Structural Versus Functional Coverage

The *functional coverage* of a specification is the fraction of the interesting behaviors that are covered by the specification. The goal of all coverage metrics is to approximate this value. A *structural* coverage metric is defined over some structural feature of the implementation – it is the fraction of the structural features that are covered by the specification.

All the coverage metrics studied by us so far are structural in nature. They evaluate the coverage of the structure of the implementation by performing mutations on the structure and verifying whether this mutation has any effect on the satisfaction of the specification. We had noted earlier that it is natural to expect that FPV coverage metrics will be structural, since any formal functional specification of the coverage goal can itself serve as the specification.

We had also noted that a structural coverage metric is meaningful if a low value of coverage implies low functional coverage. The reverse is not true – a structural FPV coverage metric may report a high value of coverage when the functional coverage of the specification is low. We shall demonstrate this fact in this section. It helps us to understand the limitations of these coverage metrics.

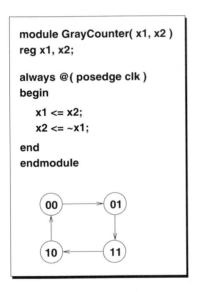

Valid implementation

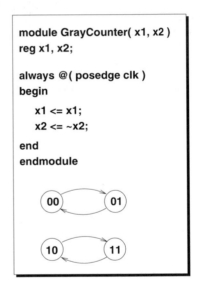

Invalid implementation

Fig. 5.3. Two-bit gray encoded counter

Let us consider a two-bit gray encoded counter. A gray encoded counter has the following property:

> *The next value of the counter differs from the present value of the counter in exactly one bit.*

Fig 5.3 shows two implementations of the counter and the corresponding FSMs. The first one is a correct implementation. The second one is an incorrect implementation since it does not reach all states, but it also satisfies the above property.

The property yields 100% state coverage in the first (correct) implementation. If we toggle any of the state bits, then the state differs from the next state in two bits or none, thereby refuting the above property.

Does this mean that the property covers the functionality of gray encoded counters? To see that this is not the case we observe that with this property, we also get 100% state coverage in the second (incorrect) implementation!

Do we need to write more properties? Yes, indeed. We need to add a property that makes sure that the counter actually counts, that is, it visits all other states before revisiting a state.

This establishes that 100% structural coverage does not necessarily mean that we have written enough properties. We get useful feedback from FPV coverage metrics mostly when the coverage is low – it is a definite indicator that more properties need to be written.

5.4 Fault-based FPV Coverage

There are two major limitations of the mutation-based approach to FPV coverage. These are:

1. *An implementation is required for coverage analysis.* We perform mutations on the given implementation and compare the truth of the property in the original implementation with the truth of the property in the mutant implementation.

2. *Coverage analysis relies on the existence of a FPV tool.* If the FPV tool runs into capacity issues, then we cannot check the truth of the specification in the implementation, and thereby cannot perform coverage analysis.

In recent times there has been a growing need for developing coverage metrics that can be used to evaluate the completeness of a specification without any implementation as the reference. It is increasingly being realized that property verification is very useful at the higher levels of the design flow, where one attempts to model the design intent in terms of formal properties. This is a challenging task, since at the highest level the validation engineer has to *identify* the essential properties from the design's architectural specification, which is usually an English document. These properties are then coded in some formal assertion language such as PSL or SVA, and is popularly called an *assertion IP* for the design.

During the development phase of an assertion IP, often the specification document (in English) is the only guideline. For example, while developing an assertion IP for a standard protocol, such as PCI XP, ARM AMBA, or IBM Coreconnect, our only reference is the protocol specification. Moreover, the assertion IP for a protocol such as ARM AMBA Bus targets not just a single implementation that uses the AMBA Bus for communication, but *all* systems that use the AMBA Bus.

What do we have at this level? We do have the design architecture that defines the interfaces of the main architectural blocks. The interfaces signals (and their directions) for each block is known at this level, but we do not have the implementation of the blocks. The protocol specification consists of properties over the interface signals of individual blocks (component properties) as well as properties over interface signals of multiple blocks (system properties). We have to develop the assertion suite based on these properties and verify whether we have written enough properties.

How can we assess the completeness of the specification at this level? Existing mutation based coverage estimation methods cannot be applied here because (a) these metrics require an existing implementation, and (b) these metrics require FPV techniques which have capacity limitations. Assertion IPs are also used in dynamic property verification (in fact, there are more users of dynamic assertion-based verification than static formal property verification), and it is just as important to verify whether the assertion IP has enough properties for dynamic verification as for static FPV. Assertion IPs for dynamic verification can be quite large, since these are not constrained by the capacity of an FPV tool. Existing coverage metrics will typically fail to analyze these specifications because of their dependence on FPV tools.

We now present a new style of assessing the coverage of a formal specification in the absence of any implementation[1]. Instead of using an implementation as a reference we use a fault model as a reference. The coverage analysis verifies whether the specification remains *consistent* in the presence of the fault – if not, then the fault is covered by the specification, otherwise the specification cannot detect the fault. We show that we need to look beyond *satisfiability* to determine whether the fault-injected specification is consistent.

What can be our fault model? Our methodology can work with any fault that can be modeled in the language used for coding the assertions. We shall choose the single stuck-at fault model over the interface signals to demonstrate

---

[1] This style of coverage analysis was first presented in [45].

the new approach – the same style and the same methodology can be used for other fault models as well.

Let us recall that all existing FPV coverage metrics are structural in nature – low structural coverage is usually associated with low functional coverage, but high structural coverage does not necessarily indicate high functional coverage. Expecting an FPV coverage metric to be a functional coverage metric leads us to the paradox where the formal coverage goal is itself the desired formal specification.

Our choice of the fault model also comes within the same paradox. If we attempt to cover all behavioral faults in our fault model, then the task of building the fault model will become as difficult as the task of building the assertion IP – in fact one task is really the dual of the other. The advantage lies in using a simple fault model that shows low coverage when the specification has gross inadequacies. Our interactions with practioners of FPV reveals that the feedback obtained from such simple metrics (that have low overheads) are very valuable during the development of an assertion IP.

We first define our structural coverage goal, and then show the use of the single stuck-at fault model to achieve this goal. We distinguish between the input signals and non-input signals while analyzing the specification, which is quite natural since input signals are controlled by the environment, and the specifications can at best specify some *assumptions* on the input behavior. Our coverage goal is as follows:

- *Non-input signals.* A specification is complete with respect to a non-input signal, s, if it covers at least some behaviors where s is required to be high, and some behaviors where s is required to be low.

- *Input signals.* A specification is complete with respect to an input signal, z, if it covers at least some behaviors that are triggered by raising z, and at least some behaviors that are triggered by lowering z.

After designing and analyzing several verification IPs, and also through interactions with validation engineers who are attempting to use FPV, we have reasons to believe that this is a reasonable and meaningful first-cut coverage goal.

In order to formally analyze a given specification against this coverage goal, we use the single stuck-at fault model. For example, to test whether the specification is complete with respect to a non-input signal, s, we first check whether the specification remains consistent in the presence of the fault, s stuck-at-0. If not, then there is some behavior covered by the specifications that conflicts with the fault – these are the behaviors where s is expected to be high. On the other hand, if the specification remains consistent in the presence of the fault, then the specification models none of the behaviors where s is expected to be high. The test for consistency with respect to s stuck-at-1 is similar, except that it looks for behaviors where s is expected to be low. Coverage with respect to input signals is treated in a slightly different way, but the essence remains the same.

An obvious criticism of our approach is that one may write properties targeting the fault model and thereby achieve a false sense of coverage. For example, if the validation engineer adds properties to indicate that every non-input must be high sometimes, and must also be low sometimes, then the proposed metrics will indicate 100% coverage with respect to the non-input signals. But it should be noted that other structural coverage metrics are also subject to the same criticism. As shown in the last section, it is possible to present an RTL, that gives 100% coverage with respect to a given metric, but does not actually implement the desired functionality. The correct approach is to write the specifications from the functional point of view (that is, without having the coverage goal or fault model in mind), and then use the structural coverage metrics to find out whether there are any gross coverage gaps. Additionally, while interpreting the coverage results for any structural coverage metric, one must keep in mind that a low coverage indicates that the specification is incomplete, but a high coverage does not necessarily guarantee functional completeness.

5.4.1 The Coverage Strategy

The key idea behind our formal coverage analysis is based on checking the realizability of a specification in the presence of a fault. The following example defines a simple toy specification which will be used to demonstrate both the methodology as well as the semantics of the proposed style of coverage.

Example 5.1. We consider the specification of a 2-way priority arbiter having the following interface:

mem-arbiter(input r_1, r_2, non-input g_1, g_2)

Let us now consider an incomplete specification for the arbiter.

1. *Request line r_1 has higher priority than request line r_2.* Suppose the design architect expresses this property as:

$$\varphi_1 : \qquad G(r_1 \wedge r_2 \Rightarrow X \ g_1)$$

which says that g_1 is favored whenever we have contention between r_1 and r_2.

2. *The grant lines are mutually exclusive.* This may be expressed as:

$$\varphi_2 : \qquad G(\neg g_1 \vee \neg g_2)$$

Let us assume for the time being that the specification consists of only the above properties, that is, $\mathcal{T} = \varphi_1 \wedge \varphi_2$. We purposefully consider the incomplete specification to demonstrate the coverage analysis in later examples.
□

The new style of coverage analysis is broadly as follows. We will model faults on the input and non-input signals and then verify whether the specification remains consistent in the presence of the fault. If not, then the specification covers the fault, otherwise we have found a coverage gap between the specification and the fault model. Faults on the non-input lines may affect the behavior of the module, while faults on the input lines may affect the *triggering* of one or more properties. Therefore we treat faults on inputs and non-inputs separately.

The next couple of sections demonstrate the above notion of fault analysis through examples, and present the formal method for coverage computation. For checking the consistency of a specification in the presence of a fault, we shall use the notion of *realizability* (see Chapter 4) of specifications.

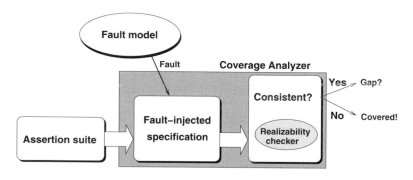

Fig. 5.4. Fault-based Coverage Analysis

5.4.2 Coverage of Faults on Non-inputs

The formal definition of coverage of faults on non-inputs is as follows.

Definition 5.2. [Coverage of faults on non-inputs:]
A stuck-at fault on a non-input signal is covered by a realizable specification,

$\mathcal{T}(\mathcal{I}, \mathcal{Z})$, *iff it is impossible to create an implementation that has the fault and yet realizes the specification.* $\square$

There are two notable issues here. Firstly, we assume that at the point of coverage analysis, we have already verified that the specification is realizable. If the specification is unrealizable, we indicate no coverage at all. Secondly, if the specification is vacuous (that is, it is always true), then the injection of any single fault will not affect its truth, and we will report zero coverage. These are the two extremes. The following example shows the typical scenario where some faults are covered, while others are not.

Example 5.3. We consider the specification presented in Example 5.1. Our coverage goal is to check whether the specification covers behaviors where each non-input is required to take each possible value.

Let us first examine whether the given specification covers any behavior where g_1 is required to be high. Indeed, the property

$$\varphi_1 : \qquad G(r_1 \wedge r_2 \Rightarrow X \ g_1)$$

covers some of the desired behaviors – g_1 is required to be high whenever r_1 and r_2 are asserted together. Let us now see how the stuck-at-fault model enables us to arrive at the same result.

While looking for the coverage of behaviors where g_1 is required to be high, we will test the realizability of the specification in the presence of a stuck-at-zero (s-a-0) fault at g_1. This is done by modeling the fault as the LTL property $G(\ \neg g_1\ )$ and then testing the realizability of the conjunction of this property with the specification. In this case, the property φ_1 cannot be realized in the presence of the fault, and hence we conclude that the specification covers some of the behaviors where g_1 is required to be high.

In many cases, the fault coverage may not be so obvious. For example, while looking for the coverage of behaviors where g_2 is required to be low, we will test the realizability of the specification in the presence of a s-a-1 fault at g_2. In the presence of this fault, the property:

$$\varphi_2 : \qquad G(\neg g_1 \vee \neg g_2)$$

can only be satisfied if g_1 is always low. But this requirement is inadmissible in conjunction with φ_1. Therefore, the specification covers some behaviors where g_2 must be low.

Finally, let us examine the coverage of behaviors where g_1 is required to be low. The fault g_1 s-a-1 is not covered by the specification, $\mathcal{T}$. For example consider an implementation of the arbiter which always asserts g_1 and never asserts g_2. Such an implementation satisfies $\mathcal{T}$. The coverage analysis discovers the fact that *there are no properties in the specification that (directly or indirectly) require* g_1 *to be de-asserted.* Likewise, the same counter-example shows that the fault g_2 s-a-0 is not covered by $\mathcal{T}$.

A coverage feedback points out the gross absence of properties covering specific types of behaviors. In this example, the coverage report will show that the specification does not cover any behavior where g_2 should be high, nor any behavior where g_1 should be low. Indications of such gross coverage gaps are extremely valuable during the early phases of writing a formal specification. The intuitionistic nature of human reasoning often causes an engineer to overlook such gaps in a formal specification.

For example, in this case on studying the coverage report, the engineer may add a new property, such as:

$$\varphi_3 : \qquad G(r_2 \wedge \neg r_1 \Rightarrow X \; g_2)$$

which requires g_2 to be asserted when r_2 is the sole request. The addition of this property into the specification directly covers the fault g_2 s-a-0, and indirectly (with φ_2) covers the fault g_1 s-a-1. $\square$

At this point we would like to point out two notable issues about our coverage analysis methodology. Firstly, the coverage metric looks for the coverage of *some* behaviors where a non-input is required to assert a specific value. It does not indicate whether *all* behaviors of that type are covered. For example, in the previous example, the coverage analysis shows that the specification covers some of the behaviors where g_1 is required to be high. It does not indicate whether there can be other scenarios, not covered by the specification, where g_1 should also be high. In our example, the cases where r_1 is the sole request are such cases, where g_1 must be asserted (in the next cycle), but are not covered by the specification. This is a normal limitation of any structural coverage metric – if we were to be able to ascertain the coverage of *all* behaviors, then we would have to specify all valid behaviors – which leads us to the paradox described in the introduction.

The second issue which warrants an explanation is: *Why do we need a realizability check in our coverage analysis?* Properties of open systems (such as RTL modules) typically specify the behavior of the module under given input scenarios. The properties do not specify that these input scenarios will actually occur. A property is *vacuously* satisfied in all runs where the inputs do not match the scenarios for which the property is written. For example, the property φ_1 in the previous example is vacuously satisfied in all runs where r_1 and r_2 never arrive together. These runs are not affected by faults on non-inputs, and thereby the property remains *satisfiable* in the presence of faults on non-inputs. For example, even though φ_1 covers some behaviors where g_1 is required to be high, it remains satisfiable (albeit vacuously) in the presence of a s-a-0 fault on g_1.

On the other hand, *realizability* has the stronger requirement that the property must be satisfied over *all* input scenarios. Therefore if a property specifies some behavior under some specific input scenarios, then the realizability check verifies the satisfiability of the specified behaviors under each of those input scenarios. For example, φ_1 is not realizable in the presence of a s-a-0 fault on g_1, since φ_1 cannot be satisfied in the presence of this fault for those input scenarios where r_1 and r_2 arrive together.

Lemma 5.4. *Let $\mathcal{T}$ be a given realizable LTL property and let z be a non-input signal. Then $\mathcal{T}$ covers the fault z s-a-0 iff $\mathcal{T} \wedge G\neg z$ is unrealizable, and $\mathcal{T}$ covers the fault z s-a-1 iff $\mathcal{T} \wedge G\, z$ is unrealizable.*

Proof: If $\mathcal{T} \wedge G\neg z$ is realizable, then there exists a possible implementation that keeps z at zero forever (since it satisfies $G\neg z$) and yet satisfies $\mathcal{T}$. Then, by the definition of the coverage, the fault z s-a-0 is not covered by $\mathcal{T}$.

If $\mathcal{T} \wedge G\neg z$ is unrealizable, then every implementation that realizes $\mathcal{T}$ must assert z sometime in the future (otherwise, the implementation would also realize $\mathcal{T} \wedge G\neg z$). Therefore the fault z s-a-0 is covered by $\mathcal{T}$. We use a similar reasoning for the fault z s-a-1. □

5.4.3 Coverage of Faults on Inputs

There is an important semantic difference between coverage of faults on inputs and those on non-inputs. We use faults on a non-input to determine whether the specification covers some behavior where the non-input must be high/low. On the other hand, we use faults on an input line to check whether the input value necessarily affects the module behavior. In other words, if a module can satisfy the specification without actually reading the given input, then we have a coverage gap, because the specification does not cover any behavior where the input line is relevant. As in the case of non-inputs, we have two types of behaviors – those that are triggered by a high value of the input, and those that are triggered by a low value of the input.

Based on the opinion of several practitioners of FPV, we provide two types of input fault coverage metrics, namely, *strong* and *weak*. The individual merits and limitations of these metrics will become apparent through the following discussion.

Definition 5.5. [Strong Input Fault Coverage:]
A stuck-at fault on an input signal is strongly covered by a realizable specification, $\mathcal{T}(\mathcal{I}, \mathcal{Z})$, unless it is possible to implement a module that realizes the specification without reading that input and assuming that it always takes the same value as the fault. □

Example 5.6. Let us again consider the enhanced specification described in Example 5.3. The specification consists of the properties:

$$\varphi_1 : \qquad G(r_1 \wedge r_2 \;\Rightarrow\; X\, g_1)$$

$$\varphi_2 : \qquad G(\neg g_1 \vee \neg g_2)$$

$$\varphi_3 : \qquad G(r_2 \wedge \neg r_1 \;\Rightarrow\; X\, g_2)$$

Our coverage goal for an input signal is to check whether the specification covers any behavior that gets triggered when the input becomes high, and whether it covers any behavior that gets triggered when the input signal becomes low. To check this, we inject an appropriate fault into the input line.

For example, let us first consider the input r_1. Indeed it is necessary for the module to read r_1 to be able to satisfy the specification. To see this, consider the case where r_2 is high at time t. If r_1 is high at time, t, the module is required to assert g_1 at $t+1$ (by φ_1), otherwise it must assert g_2 (by φ_3). It cannot assert both g_1 and g_2 (by φ_2). Therefore, it must read r_1 to make the correct decision.

To analyze this case formally, we inject the fault r_1 s-a-1 into the specification. If an implementation, J, satisfies φ_1 without reading r_1 and assumes that r_1 is always high, then that implementation must satisfy the following restriction of φ_1:

$$\varphi_1' : \qquad G(r_2 \;\Rightarrow\; X\, g_1)$$

Now consider the case where r_1 is actually low and r_2 is high at time t. Now J must assert both g_1 (by φ_1') and g_2 (by φ_3) at $t+1$. This conflicts with φ_2. In other words, the fault injected specification consisting of φ_1', φ_2 and φ_3 is unrealizable. We conclude that the input fault r_1 s-a-1 is strongly covered by the specification, and that the specification covers such behaviors that are triggered when r_1 goes high.

Does a module need to read r_2 in order to satisfy the specification? Curiously, the answer is negative! Consider an implementation that asserts g_2 (by default) in all cycles, except in those where r_1 had arrived in the previous cycle, that is, it asserts g_1 only in these remaining cycles. Such an implementation satisfies the specification without reading the input r_2 – formally, it assumes r_2 to be stuck at 1.

Our coverage analysis will show that the specification does not cover the fault r_2 s-a-1. This will mean that the specification does not cover any behavior where the module will necessarily have to consider the low value of r_2. In this example, it shows that the input r_2 is redundant. $\square$

Let $\mathcal{T}$ be a given realizable LTL specification and let v be an input signal. Let $\mathcal{T}_v^0$ denote the modified specification with v instantiated to 0, and $\mathcal{T}_v^1$ denote the modified specification with v instantiated to 1.

Lemma 5.7. *The specification $\mathcal{T}$ strongly covers the fault v s-a-0 iff $\mathcal{T} \wedge T_v^0$ is unrealizable, and $\mathcal{T}$ covers the fault v s-a-1 iff $\mathcal{T} \wedge T_v^1$ is unrealizable.*

Proof: Any implementation that satisfies $\mathcal{T}$ without reading the input v and assuming it to be 0, must satisfy T_v^0. If $\mathcal{T} \wedge T_v^0$ is unrealizable, then there can be no such implementation. By definition of strong input coverage, the fault is covered.

On the other hand, if $\mathcal{T} \wedge T_v^0$ is realizable, then there exists a realization which satisfies $\mathcal{T}$ without reading the input v and assuming it to be 0. Therefore the fault is not covered. We can use a similar reasoning for the fault v s-a-1. $\square$

Curiously, the above metrics for strong input fault coverage showed a low degree of coverage for several specifications that validation engineers thought were quite formidable. The following example explains the arguments that came up.

Example 5.8. Let us again consider our specification for the arbiter:

$$\varphi_1 : \qquad G(r_1 \wedge r_2 \;\Rightarrow\; X\, g_1)$$
$$\varphi_2 : \qquad G(\neg g_1 \vee \neg g_2)$$
$$\varphi_3 : \qquad G(r_2 \wedge \neg r_1 \;\Rightarrow\; X\, g_2)$$

Strong input fault coverage showed that the specification did not contain any property for which the module needed to read the input r_2. A module could simply satisfy the specification by parking the grant on g_2 whenever r_1 was low.

However, it also needs to be noted that the properties φ_1 and φ_3 are satisfied non-vacuously only when r_2 is high. Therefore, the specification does address behaviors that require r_2 to be high. In other words, this pointed out to the need of an additional coverage metric that would show the types of input scenarios covered, regardless of whether the corresponding behavior of the module could be satisfied by default. $\square$

The notion of *weak input fault coverage* is quite simple. Since the goal is to check whether an input non-vacuously affects the specification, we simply inject the appropriate fault into the specification and check whether it still logically implies the original specification.

Definition 5.9. [Weak Input Fault Coverage:]
A s-a-0 fault on an input signal v is weakly covered by a realizable specification, $\mathcal{T}$, unless $T_v^0 \Rightarrow \mathcal{T}$. Likewise, a s-a-1 fault on an input signal v is weakly covered by a realizable specification, $\mathcal{T}$, unless $T_v^1 \Rightarrow \mathcal{T}$. $\square$

Lemma 5.10. *If a fault is strongly covered by a realizable specification,* $\mathcal{T}$, *then it is also weakly covered by* $\mathcal{T}$.

Proof: Let the fault be v s-a-0. Then $\mathcal{T}$ strongly covers the fault iff $\mathcal{T} \wedge \mathcal{T}_v^0$ is unrealizable. Now, if $\mathcal{T}$ is realizable, then so is $\mathcal{T}_v^0$, because the input v s-a-0 is also one of the possible input scenarios.

Let us assume the contrary, namely that $\mathcal{T}$ does not weakly cover the fault. Then the following must be valid (from definition):

$$\mathcal{T}_v^0 \Rightarrow \mathcal{T}$$

This implies that the implementation that realizes $\mathcal{T}_v^0$ also realizes $\mathcal{T}$ and hence realizes $\mathcal{T} \wedge \mathcal{T}_v^0$. This is a contradiction. □

5.4.4 LTL-Covanalyzer: The Tool

The main technology that is required for implementing the proposed style of coverage analysis for formal specifications is the realizability checker. Realizability checking is a hard problem – it has been shown to be 2EXPTIME complete for LTL specifications. However, we draw attention to the fact that though existing LTL model checking algorithms have complexities that are exponential in the length of the property, the main complexity of LTL model checking lies, not in this exponential, but in handling the state explosion arising out of the product of the modules in the implementation. The fault based coverage methodology is free from this state explosion problem, since we use a fault model as the reference instead of using an implementation as the reference (as in mutation based approaches). Thereby this approach is much more scalable in practice, in spite of being doubly exponential in the length of the properties.

We used our in-house SpecChecker tool presented in Chapter 4 to perform the realizability checks. We developed two versions of the tool, namely LTL-Covanalyzer, which works on LTL specifications using SpecChecker, and Forspec-Covanalyzer, which works on specifications written in the Intel Forspec language. Forspec-Covanalyzer is semantically equivalent to LTL-Covanalyzer, but uses a different computational approach as outlined in the next section.

The tool has been successfully tested over several industrial specifications, including the ARM AMBA AHB protocol suite. This property suite consists of 36 properties, of which 22 properties are for the master interface, 9 are for the slave interface and the remaining are for the arbiter interface.

Analysis of the coverage report for the AMBA AHB specification revealed some interesting gaps in the specification. For example, the coverage report for the arbiter specification indicated that the s-a-1 fault on the non-input signal, HMASTLOCK, was not covered. This was a serious gap, since HMASTLOCK must only be asserted when a transaction needs to be locked, and not otherwise. Consider any arbiter design that incorrectly locks every transaction – such an arbiter would pass all the properties in the existing specification. In fact, the only property over the HMASTLOCK signal in the arbiter specification was:

G ((HGRANT & HREADY & HLOCK) $\Rightarrow$ X (HMASTLOCK))

The above property covers the s-a-0 fault on HMASTLOCK and therefore covers some behaviors where HMASTLOCK must be asserted, but it covers none of the behaviors where HMASTLOCK must be lowered.

We found similar gaps in most of the specifications that we analyzed. The coverage reports assume significance when we consider the fact that most of these specifications were developed by experts and had passed through several rounds of reviews. This supports the general belief that:

1. It is always possible that we have not written some important properties, and

2. It is hard for humans to find such coverage gaps without using formal methods.

5.4.5 Building it Over FPV Tools

Though our coverage analysis methodology is simple, the main challenge in building a tool for the fault based coverage approach lies in developing a realizability checker. The SpecChecker tool demonstrates that this is feasible.

Can we use the same style of coverage analysis without using a realizability checker? In this section we shall show that an approximate solution is certainly possible – a solution that uses only a satisfiability checker.

There are distinct advantages in this approach. Satisfiability checkers for most of the specification languages are built-in into existing FPV tools. Also for practical purposes an approximate coverage result is acceptable provided we consistently under-estimate the coverage by the specification.

Intuitively, the proposed approximation to a realizability check works by adding input constraints to the specification that force non-vacuous interpretations of the properties, and then by testing the satisfiability of the modified specification.

Example 5.11. Let us again consider the following LTL property for our arbiter from the earlier examples:

$$\varphi_1 : \qquad G(r_1 \wedge r_2 \;\Rightarrow\; X\; g_1)$$

The property is vacuously satisfied if r_1 and r_2 are never asserted together. In such cases the value of g_1 may be treated as a don't care. The property is therefore satisfiable even in the presence of a s-a-0 fault in g_1. To generate the witness, the satisfiability checker finds input scenarios other than those for which g_1 is required to be asserted. Our aim is to add input constraints to block such witnesses.

Suppose we add the following input constraint into the specification:

$$\psi_1 : \qquad GF(r_1 \wedge r_2)$$

The input constraint specifies that those scenarios where r_1 and r_2 arrive together, will occur infinitely often.

The modified specification consisting of φ_1 and ψ_1 is not satisfiable in the presence of a s-a-0 fault in g_1, because ψ_1 prevents the satisfiability checker from producing only vacuous interpretations of φ_1. In other words ψ_1 enforces those scenarios where φ_1 requires g_1 to be asserted. □

Formally, a property is *vacuously satisfied* on a run if the instantiations of the input signals along the run is sufficient to satisfy the property (regardless of the values of the non-inputs). If we are able to find any input scenario for which the fault-injected specification becomes unsatisfiable, then we can safely conclude that the fault-injected specification is unrealizable as well, and that the fault is covered. This follows from the fact that a realizable specification must necessarily be satisfiable under all input scenarios. We add constraints on the input space to restrict the search to *non-vacuous* runs only.

It is important to note that:

- Unsatisfiability under any input constraint is *sufficient* to establish the un-realizability of a specification.

- If the specification remains satisfiable under our input constraints, we cannot decide whether the specification is realizable. This is because of our inability to cover all input scenarios in general.

Therefore, the coverage results are an approximation of the actual coverage. However our estimate is conservative in the sense that:

- If a fault is found to be covered by our analysis, then it is truly covered, since we have a proof for the un-realizability of the fault injected specification.

- If a fault cannot be covered by our analysis, then it may or may not be covered in reality.

For the second case, false negatives will be rare in practice if we can automatically identify the input scenarios for which the property has a non-vacuous interpretation. This is a non-trivial task because existing property specification languages do not have any syntactic separation between the specification of the behavior and the input scenario under which that behavior is mandatory.

One possible option is to request the validation engineer to separately provide the input restrictions under which the property is expected to be satisfied non-vacuously. From a practical point of view this is both a good and bad option, good – because validation engineers typically know the scenarios for which a property has been written, and bad – because our goal is to obtain a simple first-cut metric that requires no user intervention. The latter motivated us to develop an approximate strategy for automatically identifying the input scenarios that *trigger* non-vacuous interpretations of a temporal property. We believe that this strategy will find use in other FPV applications as well.

Example 5.12. Let us again consider the two-input arbiter of our previous examples. Suppose the specification contains the following property:

Whenever r_1 is asserted and r_2 is low, the arbiter asserts the grant g_1 sometime in future, unless r_1 is lowered in between.

This requirement can be expressed by the following LTL properties:.

$$\varphi: \qquad G(r_1 \wedge \neg r_2 \ \Rightarrow \ X \ (r_1 U(g_1 \vee \neg r_1)))$$

Note that φ is sufficient to cover the stuck-at-0 fault on the non-input line g_1. In order to enforce the case where the arbiter has to assert g_1, the environment must lower r_2, and keep r_1 asserted. This is the input scenario that we want for coverage analysis. Our methodology will produce the following constraint that has the desired effect:

$$I_\varphi: \qquad F(r_1 \wedge \neg r_2 \ \wedge X \ (G(r_1)))$$

$\square$

Our approach is to eliminate non-inputs from the properties until we are left with a constraint over the inputs only. The algorithm for eliminating non-inputs uses the following partial order among properties.

Definition 5.13. [Strong and weak properties:]
A property $\mathcal{F}_1$ is stronger than a property $\mathcal{F}_2$ iff $\mathcal{F}_1 \Rightarrow \mathcal{F}_2$ and $\mathcal{F}_2 \not\Rightarrow \mathcal{F}_1$. We also say that $\mathcal{F}_2$ is weaker than $\mathcal{F}_1$. $\square$

Given a property, φ, defined over input variables $\mathcal{I}$, and non-input variables $\mathcal{O}$, we generate a formula S_φ that is stronger than φ and is defined only over $\mathcal{I}$. Since S_φ is stronger than φ and is free from non-inputs, it follows that S_φ describes input scenarios that make φ vacuously true. Therefore, we restrict the input space by $\neg S_\varphi$, which covers all non-vacuous runs.

Example 5.14. Consider the property:

$$\varphi \; : \; G[\, r_1 \; \Rightarrow \; X \; g_1 \,]$$

Then $S_\varphi = G(\neg r_1)$ and the constraint that restricts the input space to non-vacuous runs is $\neg S_\varphi = F\, r_1$. $\square$

To find the substitution of the non-inputs that gives us the desired input constraint for a property φ, we use the following function, $\mathcal{S}(\varphi)$. The function returns a set of substitutions, of the non-inputs that generates a property S_φ over $\mathcal{I}$ which is stronger than φ. The function is recursively defined as follows. We convert the property to *negation normal form* before applying this function.

Function 1 $\mathcal{S}(\varphi)$

case $\varphi = o$ *(non-input)*	*return* $o \leftarrow 0$.
case $\varphi = i$ *(input)*	*return NULL.*
case $\varphi = (i \lor o)$	*return* $o \leftarrow 0$.
case $\varphi = (i \land o)$	*return NULL.*
case $\varphi = (\psi \; U \; \phi)$	*return* $\mathcal{S}(\psi) \cup \mathcal{S}(\phi)$
case $\varphi = (\psi \lor \phi)$	*return* $\mathcal{S}(\psi) \cup \mathcal{S}(\phi)$
case $\varphi = (\psi \land \phi)$	*return* $\mathcal{S}(\psi) \cup \mathcal{S}(\phi)$
case $\varphi = X(\psi)$ *or* $\varphi = G(\psi)$ *or* $\varphi = F(\psi)$	*return* $\mathcal{S}(\psi)$

Example 5.15. Let us apply the above function on the property of Example 5.12:

$$\varphi : \qquad G(r_1 \land \neg r_2 \; \Rightarrow \; X \; (r_1 U(g_1 \lor \neg r_1)))$$

Converting to negation normal form, we have:

$$\varphi : \qquad G(\neg r_1 \lor r_2 \lor X \; (r_1 U(g_1 \lor \neg r_1)))$$

$\mathcal{S}(\varphi)$ will produce the substitution $g_1 \leftarrow 0$. This substitution will generate the property:

$$\varphi' : \qquad G(\neg r_1 \vee r_2 \vee X \ (r_1 U \neg r_1))$$

which is the same as:

$$\varphi' : \qquad G(\neg r_1 \vee r_2 \vee X \ F \ \neg r_1)$$

Negating this property gives us the desired input constraint:

$$I_\varphi : \qquad F(r_1 \wedge \neg r_2 \ \wedge X \ (G(r_1)))$$

$\square$

Since a property may have a non-input in a positive literal as well as a negative literal, $\mathcal{S}(\varphi)$ may return conflicting substitutions for such non-inputs. In such cases, we retain both substitutions and perform coverage analysis using each conflicting constraint separately. Since a realizable property must be satisfiable under all input restrictions, there is no harm in performing coverage analysis under different input constraints.

Typically, conflicting substitutions (if they exist) are meaningful. For example, consider the following LTL formula for our arbiter:

$$\varphi = G((r_1 \Rightarrow X g_1) \wedge (r_1 \Leftarrow X g_1))$$

$\mathcal{S}(\varphi)$ generates both $g_1 \leftarrow 0$ and $g_1 \leftarrow 1$. The restriction $g_1 \leftarrow 0$ generates the input constraint, $F \ r_1$, which is required to cover the stuck-at-0 fault on g_1. The restriction $g_1 \leftarrow 1$ generates the input constraint, $F \ \neg r_1$, which is required to cover the stuck-at-1 fault on g_1.

We have integrated the constraint generation algorithm into the CovAnalyzer tool and used an in-house LTL satisfiability checker to approximate the realizability checks. We also used a similar approach in our Forspec based CovAnalyzer.

The time required for approximate coverage analysis was found to be significantly less than the time required by the exact analysis (using the realizability checker of SpecChecker).

5.4.6 Other Fault Models

The fault based coverage approach can be used with other fault models as well, provided that we can express the faults as properties in the specification language and inject them into the specification. We outline three interesting variants:

1. *Multiple stuck-at faults.* This is a straight-forward extension, but has some interesting benefits. For example, consider the property which says – *whenever there is a pending request, r_1 or r_2, the arbiter will not waste any*

cycle, that is, one of the grant lines g_1 or g_2 must be asserted in the next cycle. We can write this property in LTL as:

$$G[\, r_1 \ \lor \ r_2 \ \Rightarrow \ X(\, g_1 \ \lor \ g_2 \,)\,]$$

Single stuck-at faults at g_1 or g_2 are not covered by this property. However, the property covers the multiple stuck-at fault where both g_1 and g_2 are stuck at 0.

2. *Counter-example faults.* Validation engineers sometimes list a cross section of the counter-example scenarios that are expected to be detected by the specification. We can model such counter-examples as properties – which then constitutes our fault model. The analysis methodology is similar. For example, in order to verify whether the specification guarantees mutual exclusion between three grant lines of an arbiter, we can add three counter-example scenarios where two of the lines are high at the same time. Each of these scenarios are covered if the specification becomes inconsistent in the presence of each of these faults.

3. *Short faults.* Sometimes the specification may contain bit vectors. Large bit vectors lead to serious capacity issues in FPV tools. A short fault joins two or more bits into a single bit. If such a fault does not affect the consistency of the specification, then it indicates that the specification does not distinguish between these bits. This allows the validation engineer to perform bit-scaling (that is, to cut down the bit vectors) before using the FPV tool.

We believe that many other benefits of the new style will emerge as it gets adopted in practice.

5.5 Concluding Remarks

FPV coverage is one of the most hot areas in formal verification today. As we accept the fact that FPV is not a stand-alone validation technology (at least not yet), we have to answer the question – *What have I verified?*

Currently our main interest in FPV coverage is to determine whether the validation engineer has written enough properties to cover *those behaviors that she intended to cover*. We *do not* intend to cover the entire functionality of a design through formal properties. In the long run FPV coverage metrics will have to gel with simulation coverage metrics, that is, we must be able to read FPV coverage results and specify which behaviors should be specifically targeted by simulation. Methodologies for developing an integrated test plan involving both simulation and FPV will rely to a large extent on FPV coverage metrics.

In the next chapter we present a new paradigm for formal property verification using the notion of FPV coverage. In this approach we compare two specifications at two levels of the design hierarchy and find out whether the lower level specification (having more details) covers those behaviors for which the higher level specification was written. This facilitates the validation engineer to decompose the formal specification into specifications of individual components, and helps her to get around the state explosion problem.

5.6 Bibliographic Notes

There is a considerable volume of recent literature on simulation-based coverage metrics [86, 103]. This includes recent coverage metrics such as *assertion coverage* [103] and *mutation coverage* [24, 109].

Coverage metrics for formal property verification were first proposed in [67]. At the same time, a metric for comparing an FSM and a reduced tableau for the formal specification was proposed in [72]. Since then, there has been a considerable amount of research on mutation based approaches to FPV coverage [31, 32, 33]. A comparison between simulation and FPV coverage metrics, an adaptation of simulation coverage metrics in an FPV setting, and the notion of vacuity coverage are presented in [34].

The use of a fault model in mutation based coverage analysis was first presented in [57].

The notion of evaluating the coverage of a formal specification by checking its consistency in the presence of faults chosen from a high-level fault model was presented in [45]. The main benefits of this new style of coverage analysis is that the approach can be used even when the implementation is not yet available, or when the implementation is available but beyond the capacity limits of FPV tools.

Design Intent Coverage

What is the most serious limitation of existing FPV technology? The answer is undisputed today – *capacity.*

FPV technology is poised at a peculiar position. It is well understood that there are several important advantages in using formal languages to specify the design intent at the high level (say, at the architectural level), but it is hard to verify whether the RTL satisfies these properties because most of these properties involve multiple architectural blocks of non-trivial size, and the FPV tool runs into capacity issues. We are able to use FPV technology at the unit level, but at that level the benefits are also limited.

Today most of the innovations in design optimization are happening at the architectural level – RTL synthesis has been largely standardized, leaving little room for further optimizations. Making the right design decisions at the architectural level dictates the overall performance of large digital designs, such as processors, DSPs, and memories. It is very important to ascertain that these architectural decisions are properly implemented in the RTL.

With the new languages for formal property specification, it is possible to express the key architectural decisions formally. These properties define the *architectural intent* of the design. There is a significant advantage in expressing the design intent formally. The existing practice of expressing the architectural intent through a (English) specification document leaves room for ambiguity. The gap between the architect's specified intent and the designer's perception of that intent is one of the main sources of logical bugs in the RTL.

Though microarchitects are notoriously resistant towards the use of formal properties for expressing the design intent, this is not the main technical challenge in design intent verification. The more intriguing problem is – *How can we verify whether the RTL satisfies the formal architectural intent?* Existing

FPV technology is unable to solve this problem because the existing tools do not have adequate capacity for verifying architectural properties over multiple blocks of non-trivial size.

We believe that new formal methods are needed for addressing the problem of design intent verification. Through recent collaborative research between our group and Intel, we have come up with a new paradigm for formal property verification, which attempts to solve the problem through FPV coverage. The new paradigm is called *design intent coverage*[1].

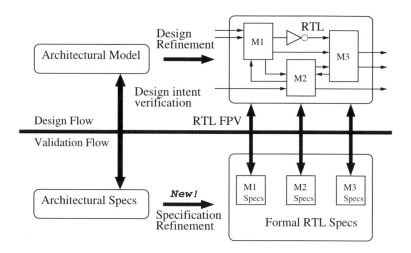

Fig. 6.1. Design Intent Coverage

The notion of design intent coverage is intuitive and beautifully simple (see Fig 6.1).

1. The key architectural properties are specified using a formal property specification language. This forms the *golden* specification, $\mathcal{A}$.

2. We identify the functional blocks in the design implementation. These are black boxes showing the relevant interface signals.

3. We identify the properties that each functional block must satisfy to achieve its role in the design. The important thing is to write these block properties *before* writing the RTL for these blocks, and *not* after (as is the common practice today). The properties of the RTL blocks taken together is called the RTL specs, $\mathcal{R}$.

---

[1] The notion of design intent coverage was introduced in [14] and further developed in [44] and [15].

4. We use new formal methods to check whether the RTL specs, $\mathcal{R}$, covers the architectural specs, $\mathcal{A}$. We refer to this question as the *primary coverage question*.

All these steps take place *before* writing any RTL code. If we find that the RTL specs cover the architectural specs, then we have achieved two important objectives, namely:

1. *Capacity.* We have reduced the complexity of the task of verifying the architectural properties. If we now verify each individual RTL block with its own RTL properties, we will be able to guarantee that the architectural properties are satisfied on the whole design. If the designer further refines a possibly large RTL block into a set of RTL sub-blocks, we will adopt the same approach recursively to refine the specs of that block into the specs of its sub-blocks.

2. *Completeness.* We have proved that enough RTL properties are written to cover the design intent specification. This is the true intent of functional coverage, as opposed to the existing structural coverage metrics presented in the last chapter.

On the other hand, if we find that an architectural property is not covered by the RTL specs, then we have several ways of demonstrating the coverage gap. One way is to produce a run that refutes the architectural property, but satisfies all the RTL properties. A single run does not show the whole of the coverage gap, and enumerating all runs that serve as counterexamples is clearly impractical. We therefore present the coverage gap in terms of: (a) RTL properties, and (b) architectural sub-properties derived out of the original architectural property. In both cases, the properties must be presented in a form that is syntactically close and visually comparable with the original properties. The algorithms presented in this chapter use the syntactic structure (grammar) of the property specification language to achieve this objective. These algorithms are integrated into our tool, SpecMatcher, which is also presented in this chapter.

We believe that the design intent coverage paradigm will significantly change the way in which FPV is used within the pre-silicon design validation flow. Since this methodology is used before RTL implementation, it enables the design architect to formally refine the RTL specifications to the extent that it formally guarantees conformance with respect to the architectural intent *even before a single line of RTL is written.* In the past, the inability to produce this kind of guarantee has led to the detection of architectural violations later in the design flow, thereby creating some expensive cycles in the design flow. A second advantage of the proposed methodology is that it scales to large designs, since we continue to recursively adopt this approach along the module hierarchy until the modules are small enough to be accepted by existing FPV tools.

A third very important benefit of this approach is that it enables validation reuse. Since validation accounts for more than 70% of the design cycle, design reuse will have only marginal benefits unless we are able to reuse the validation effort. In future, we expect reusable blocks to come with a set of properties that are guaranteed (pre-verified) for that block. Design intent coverage enables a designer to reuse one or more modules in a given context by checking that the properties guaranteed by these (and other) modules cover the design intent.

6.1 An Introductory Example

In this section we will demonstrate the notion of design intent coverage through a small example. We will show how we can verify the architectural intent through specification coverage *before* writing the RTL of the design.

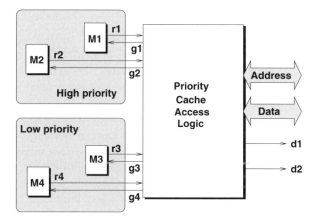

Fig. 6.2. Priority Cache Access

6.1.1 Priority Cache Access: *Architectural Specs*

Our goal is to develop a cache memory access logic that allows storage and retrieval by four different devices, M_1, M_2, M_3 and M_4. The architectural interface of this block is shown in Fig 6.2. The description of the interface signals are as follows.

1. *Request lines.* The block receives four request lines, namely r_1, r_2, r_3 and r_4 respectively from M_1, M_2, M_3 and M_4.

2. *Grant lines.* The block arbitrates between the two high-priority request lines to assert g_1 or g_2, and between the low-priority request lines to assert g_3 or g_4.

3. *Cache ready lines.* The line d_1 indicates that the page requested by the high-priority device is available in the cache. If there is a cache hit, then d_1 is asserted within two cycles. Otherwise, d_1 is asserted after the page has been fetched from memory. Similarly the line d_2 is asserted when the page requested by M_3 or M_4 becomes available in the cache.

4. *Data and address lines.* The data and address lines are used to float the cache address and store / retrieve the data.

There are many architectural requirements of this block, but for simplicity we will choose the following property, which defines the notion of priority.

M_1 and M_2 have higher priority over M_3 and M_4. A page requested by M_1 or M_2 is either served two cycles after the request (when we have a cache hit), or is served in some future cycle (when we have a cache miss). In the later case, *no page should be served to the low priority devices in the intermediate cycles.*

The above property can be expressed in Linear Temporal Logic (LTL) as:

$$A1: \qquad G[\, r_1 \lor r_2 \ \Rightarrow\ XX[\, d_1 \lor X(\, \neg d_2\ U\ d_1\,)\,]\,]$$

At this point we do not yet have any implementation. Architectural properties such as these are written by interpreting the architectural specification document.

We may also have *assume* properties at the architectural level. These properties specify the assumptions made about the environment behavior. For example, we make the following assumption for our priority cache:

A request line is not lowered until it is served.

We may formally express this assumption using the following *assume* properties.

$$G[\, r_1 \ \Rightarrow\ X[\, d_1 \lor r_1\,]\,]$$
$$G[\, r_2 \ \Rightarrow\ X[\, d_1 \lor r_2\,]\,]$$
$$G[\, r_3 \ \Rightarrow\ X[\, d_2 \lor r_3\,]\,]$$
$$G[\, r_4 \ \Rightarrow\ X[\, d_2 \lor r_4\,]\,]$$

6.1.2 Is My Implementation Plan Correct?

The first step towards implementing our priority cache access logic is to identify the main design blocks and their functionality. The functionality of the
blocks and their interconnection defines the architecture of the design. Several
different architectures may be conceived for implementing the same design intent, but we must verify whether the chosen architecture is correct, that is,
whether the functionalities of the architectural blocks and the way they are
connected actually realizes the design intent. The main challenge is in performing this verification task *before* creating the RTL, so that we do not waste
design effort on a flawed architecture.

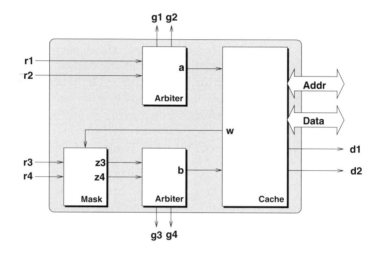

Fig. 6.3. A flawed architecture

Fig 6.3 shows one possible architecture for the priority cache. The block
definitions are as follows:

1. *Arbiters.* Two arbiters are used – one for arbitrating between the high
   priority devices (request lines, r_1 and r_2), and the other for arbitrating
   between the low priority devices (request lines, r_3 and r_4). The arbiter
   asserts the line, a (respectively b), one cycle after at least one among its
   request lines goes high.

2. *Mask.* This is a combinational logic that masks out the low priority request
   lines when w is high.

3. *Cache.* This is the cache block, including the page fetch logic. When there
   is a cache miss on a high-priority request, then the signal w is raised to
   mask the low priority request lines until the high-priority request is served.

There is a gap of at least two cycles between a high-priority request and its service – one cycle is lost in arbitration, and one cycle is lost in searching for the page in the cache. If there is a cache miss, then more cycles are needed to load the page from memory. A low-priority request may wait for more than two cycles, even when the page is available in the memory – this may happen if a high-priority request is pending, and consequently the low priority requests are masked.

What role should each architectural block (arbiters, mask, cache) play, so that the design satisfies the architectural property, A1? Today there is neither any formal way to answer this question, nor any formal way to verify whether any explanation given by the design architect is correct.

How is this problem addressed today? Two of the most common approaches are:

1. Write the RTL for the arbiters, the mask, and the cache, and then attempt to verify the property $A1$ on the whole design. This is the common practice today, and the problems are:

   a) The FPV tool cannot handle such architectural properties due to capacity limitations. In our example, the cache block can be really large, and beyond the scope of most FPV tools.

   b) Dynamic property verification (during simulation) can miss out important corner case scenarios. We shall show that the architecture shown in Fig 6.3 fails on one such corner case.

   Even if we find the error, we will find it late into the design flow, when the RTL has been written. At this stage, redesigning the architecture is a disaster.

2. The design architect is asked – *What properties do we need to prove on the component blocks such that the whole design satisfies A1?* The architect specifies some properties and explains (in English) why she expects that these properties will guarantee that the overall design satisfies $A1$. This approach is followed by many practitioners of FPV, when they have to prove a property over a large design having many components. The drawback of this approach is that the design architect's argument may be flawed, and the violations may be subtle. We need to formally ascertain whether the argument is correct. And we need to do this *before* spending time on developing the RTL.

There is a distinct advantage in the second approach. We, human beings, are very clever in eliminating unnecessary details from what we perceive. For example, if the design architect is asked – *What role do you expect the cache block to play in satisfying the architectural property, A1?* – she will more often than not produce a property that eliminates most of the irrelevant parts of the cache block specification. Truly enough, we do not need the full functionality of the cache block to prove $A1$ on the design.

Suppose the architect comes up with the following properties over the architectural blocks, which she thinks, will help in proving $A1$ on the whole design.

Arbiter properties:

$R1$: $G[\, r_1 \ \lor \ r_2 \ \Leftrightarrow \ Xa \,]$

$R2$: $G[\, z_3 \ \lor \ z_4 \ \Leftrightarrow \ Xb \,]$

Cache block properties:

$R3$: $G[\, a \ \Rightarrow \ X[\, w \ U \ d_1 \,] \,]$

$R4$: $G[\, \neg b \ \Rightarrow \ \neg X \ d_2 \,]$

Mask properties:

$R5$: $G[\, r_3 \ \land \ \neg w \ \Leftrightarrow \ z_3 \,]$

$R6$: $G[\, r_4 \ \land \ \neg w \ \Leftrightarrow \ z_4 \,]$

The arbiter properties define that the signal a (or b) is asserted when there is at least one pending request. The first cache block property, $R3$, specifies that the masking signal w is asserted until the high priority request (indicated by a) is served (indicated by raising d_1). The second cache block property, $R4$, specifies that d_2 is not asserted when there are no unmasked pending low priority requests (indicated by b). The mask properties express the masking logic – when w is high, both z_3 and z_4 are low, otherwise they have the same value as r_3 and r_4.

The architect believes that if (a) the mask satisfies $R5$ and $R6$, (b) the cache satisfies $R3$ and $R4$ and (c) the arbiters satisfy $R1$ and $R2$ respectively, then the whole design is guaranteed to satisfy $A1$.

How can we ascertain whether she is correct? We need to verify whether any behavior is possible that satisfies $R1, \ldots, R6$, but refutes $A1$. If so, then there exists some implementation (containing such behaviors) where the respective blocks satisfy $R1, \ldots, R6$, *but the whole design does not satisfy $A1$.*

Is there such a problem in our architecture? Unfortunately, yes. Fig 6.4 shows a counter-example trace. The low priority request, r_3, comes after the high priority request, r_1, but d_2 is asserted before d_1, violating the architectural property $A1$. None of the properties, $R1, \ldots, R6$, fail on this trace.

Where is the gap in the architect's argument? The problem is that there is a one cycle delay between the assertion of a and the assertion of w, when there is a high-priority cache miss. In the corner case shown above, $z3$ goes high in that gap (time $t+1$), thereby enabling the low priority request to pass

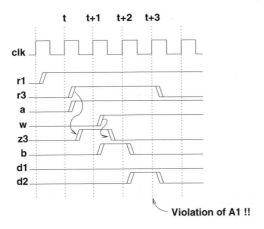

Fig. 6.4. Counter-example trace

into the cache. If there is a cache hit for the low priority request, then the page becomes available in the next cycle itself, which results in the violation of the architectural property $A1$.

6.1.3 The Correct Architecture

Fig 6.5 shows another architecture for the priority access logic. The cache block and the arbiter blocks have the same functionality, but the mask block has been moved closer to the cache block – instead of masking the low priority request, it now masks the output of the low-priority arbiter.

We now have a new set of properties over the architectural blocks. These are:

Arbiter properties:
$Q1$: $G[\, r_1 \;\lor\; r_2 \;\Leftrightarrow\; X a \,]$
$Q2$: $G[\, r_3 \;\lor\; r_4 \;\Leftrightarrow\; X b \,]$

Cache block properties:
$Q3$: $G[\, a \;\Rightarrow\; X[\, w \; U \; d_1 \,]\,]$
$Q4$: $G[\, \neg z \;\Rightarrow\; \neg X \; d_2 \,]$

Mask properties:
$Q5$: $G[\, b \;\land\; \neg w \;\Leftrightarrow\; z \,]$

The properties for the arbiters and the cache block are exactly as before. The mask property is new.

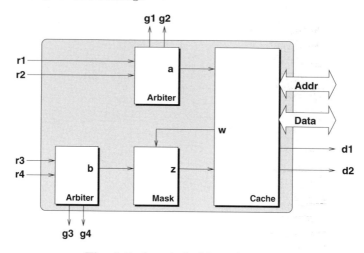

Fig. 6.5. An admissible architecture

The problem with the earlier architecture is not present in this architecture. Since the mask now applies to the output of the low priority arbiter, z will never go high when there is a cache miss for a high priority request. This ensures that d_2 is never asserted when there is a pending high priority request at the cache block.

6.1.4 How Does Design Intent Coverage Help?

The architecture shown in Fig 6.3 has a flaw – our implementation plan was incorrect. The architecture shown in Fig 6.5 is correct. *How can we formally verify these facts?*

Design intent coverage is a formal method for comparing two specifications and finding out whether one covers the other.

1. For the case shown in Fig 6.5, the design intent coverage methodology will check whether the specification consisting of the properties, $Q1, \ldots, Q5$, *covers* the architectural property $A1$. It will report that this is indeed the case. This will confirm the architect's intuition that the architectural plan is correct, and that it suffices to verify $Q1, \ldots, Q5$ on the respective blocks in order to guarantee the architectural property $A1$. For large blocks, this approach of manual decomposition and automatic coverage analysis can be used recursively, until the block sizes are small enough to be accepted by FPV tools.

2. For the case shown in Fig 6.3, the design intent coverage methodology will check whether the specification consisting of the properties, $R1, \ldots, R6$,

covers $A1$. In this case it will find that there is a gap between the specifications. On finding the gap it can do several things:

a) It can produce a counter-example trace (as shown in Fig 6.4).

b) It can produce an executable (automaton) that spins off counter-example runs.

c) It can produce a counter-example property that covers the entire gap between the two specifications.

The first two do not show the entire coverage gap, hence our focus will be on the third approach, where the missing properties are automatically detected.

The design intent coverage framework aids the validation engineer in developing a feasible formal verification plan by systematically decomposing the specifications of large design blocks into the specifications of their component modules. The formal methods used in this framework help in verifying the soundness and completeness of the decomposition, and in finding out the gaps in the decomposition. The methods are significantly scalable since we work only over specifications. We believe that the future of FPV lies in the appropriate adoption of this new paradigm.

6.2 The Formal Problem

Design intent coverage essentially compares a high level specification, A, with a low level specification, R, and determines whether R covers A, that is, whether every invalid scenario for A is also an invalid scenario for R, so that no bug detected by A is missed by R. The converse is not true – some behaviors may be invalid for R but valid for A. This is normal, since implementation specific constraints get added into the design as we go down the design flow. For example, the ARM AMBA Bus specification does not specify the exact arbitration policy, but any implementation must necessarily decide on the exact arbitration policy, thereby adding new design constraints.

To distinguish between the high level specification, A, and the low level specification, R, we shall refer to A as the *architectural intent*, and R as the *RTL specification*. This terminology is used only for the ease of presentation, and should not be viewed as a restriction on the domain of application of this technology.

We shall present the formal methods for design intent coverage using Linear Temporal Logic (LTL) as the specification language. Our in-house intent coverage tool, SpecMatcher, is also based on LTL. The main inputs to the tool are:

1. The *architectural intent* $\mathcal{A}$ as a set of LTL properties over a set, $\mathcal{AP}_{\mathcal{A}}$, of Boolean signals, and

2. The *RTL specification* $\mathcal{R}$ as another set of LTL properties over a set, $\mathcal{AP}_{\mathcal{R}}$, of Boolean signals.

We shall also use $\mathcal{A}$ to denote the conjunction of the properties in the architectural intent, and $\mathcal{R}$ to denote the conjunction of the properties in the RTL specification.

Assumption 1 *Throughout this chapter we assume that $\mathcal{AP}_{\mathcal{A}} \subseteq \mathcal{AP}_{\mathcal{R}}$.*

The above assumption essentially means that the low level specification has the same names for their signals as the corresponding ones in the high level specification. The RTL specification can have other signals in addition to these. Typically this is not a restrictive assumption within the design hierarchy, since it is generally considered a good practice for designers at a lower level of the design hierarchy to inherit the interface signal names from the previous level of hierarchy.

Given a specification, we define a *state* as a valuation of the signals used in the specification. A *run* is an infinite sequence of states.

Definition 6.1. [Coverage Definition:]
The RTL specification covers the architectural intent iff there exists no run that refutes one or more properties of the architectural intent but does not refute any property of the RTL specification. □

Our coverage problem is as follows:

- To determine whether the RTL specification covers the architectural intent, and

- If the answer to the previous question is *no*, then to determine a set of additional temporal properties that represent the coverage gap (that is, these properties together with the RTL specification succeed in covering the architectural intent).

The following theorem shows us a way to answer the first question.

Theorem 6.2. *The RTL specification, $\mathcal{R}$, covers the architectural intent $\mathcal{A}$, iff the temporal property $\mathcal{R} \Rightarrow \mathcal{A}$ is valid.*

Proof: If $\mathcal{R} \Rightarrow \mathcal{A}$, then we have $\neg\mathcal{A} \Rightarrow \neg\mathcal{R}$. Therefore any run that refutes the architectural intent also refutes the RTL specification.

To prove the opposite direction, we will use contrapositive. Suppose that $\mathcal{R} \Rightarrow \mathcal{A}$ is not valid. That means, there exists a run that refutes the architectural intent but satisfies the RTL specification. Hence, by definition, the RTL specification does not cover the architectural intent. $\square$

The theorem shows that the primary coverage question can be answered by testing the validity of $\mathcal{R} \Rightarrow \mathcal{A}$. Most model checking tools for LTL and its derivatives already have the capability of performing validity (or satisfiability) checks on temporal specifications, and can therefore be used to answer our primary coverage question. Note that the complexity of LTL model checking coincides with that of checking the satisfiability of LTL formulas (both being PSPACE-complete), but since our coverage question does not involve the RTL, the proposed approach scales to much larger designs.

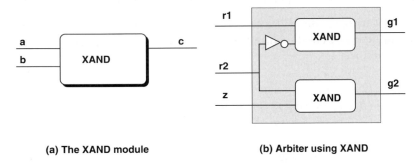

(a) The XAND module   (b) Arbiter using XAND

Fig. 6.6. A toy arbiter

We shall use a running example to demonstrate the formal methods for design intent coverage. For ease of presentation we choose a simple toy example.

Example 6.3. Let us consider the design of an arbiter that arbitrates between two request lines r_1 and r_2 from two master devices. Let the corresponding grant lines to the master devices be g_1 and g_2. The arbiter also receives an input z from a slave device, that remains high as long as the slave device is *ready*.

The arbiter specification requires us to treat r_2 as a high-priority request. Whenever r_2 is asserted and the slave is ready (that is, z is high), the arbiter must give the grant, g_2 in the next cycle, and continue to assert g_2 as long as r_2 remains asserted. When r_2 is not high, the arbiter parks the grant on g_1 regardless of whether r_1 is asserted. We are further given, that the request

r_2 is fair in the sense that it is de-asserted infinitely often (enabling g_1 to be asserted infinitely often).

The above architectural intent may be expressed in LTL as follows:

A_1: $G\ F(\ \neg r_2\ )$
A_2: $G(\ (\ r_2\ \wedge\ z\ )\ \Rightarrow\ X(\ g_2\ U\ \neg r_2\ )\ )$
A_3: $G(\ (\ \neg r_2\ )\ \Rightarrow\ X\ g_1\ )$

Let us now consider an implementation of the arbiter using a component called XAND, as shown in Fig 6.6. The specification of the module XAND is as follows:

R_1': $G(\ (\ a \wedge\ b\ )\ \Rightarrow X\ c\ )\ )$

It may be noted that we do not require the internal implementation of the RTL module XAND. Property R_1' is part of the RTL specification for XAND.

Substituting the signal names of the instances of XAND in Fig 1(b) with r_1, r_2, g_1, g_2 and z, and adding the fairness property on r_2, we have the RTL specification as:

R_1: $G\ F(\ \neg r_2\ )$
R_2: $G(\ (\ r_1\ \wedge\ \neg r_2\ )\ \Rightarrow\ X\ g_1\ )$
R_3: $G(\ (\ r_2\ \wedge\ z\ )\ \Rightarrow\ X\ g_2\ )$

The first property is the same fairness constraint as in the architectural intent. The second property says if r_1 is asserted and r_2 is de-asserted then g_1 is asserted in the next cycle. The third property states that whenever r_2 and z are asserted together, then g_2 is asserted in the next cycle.

Our primary coverage problem is to determine whether $(R_1 \wedge R_2 \wedge R_3) \Rightarrow (A_1 \wedge A_2 \wedge A_3)$ is valid. In this case the answer is negative. It is clear that A_1 is implied by the RTL specification, but we can see that neither A_2 nor A_3 is covered by the properties in the RTL specification.

For example whenever we have a scenario where both r_1 and r_2 are low, the architectural intent requires g_1 to be asserted, but the RTL specification does not have this requirement. This shows that A_3 is not covered.

Also, consider those scenarios where r_2 and z are asserted together, but z de-asserts before r_2 (that is, the slave becomes unavailable before the transfer completes). In these scenarios, the architectural intent requires g_2 to remain high as long as r_2 remains high (property A_2), but the RTL specification does not guarantee this. $\square$

6.2.1 Where is the Coverage Gap?

Theorem 6.2 shows that the primary coverage question, that is, whether the RTL specification, $\mathcal{R}$, covers the architectural intent, $\mathcal{A}$, can be answered by checking the validity of the implication $\mathcal{R} \Rightarrow \mathcal{A}$. If the implication is not valid, then we know that there is a gap between the RTL specification and the architectural intent, but *how do we find out the gap?*

As we shall show in this section, it is not hard to compute the coverage gap between two temporal specifications and specify a property that theoretically represents the coverage gap. The main challenge is in presenting the new property in a form that is syntactically similar and visually comparable with the original specification, so that the validation engineer is able to visually examine the new property and realize the set of architectural behaviors that have not been covered by the RTL specification. Let us first see how the coverage gap can be computed.

Example 6.4. Let us consider the coverage of the property, A_3 of Example 6.3 by the RTL specification. We have already established that A_3 is not covered. However this information does not accurately point out the coverage gap between A_3 and the RTL specification. Specifically the coverage gap lies only for those scenarios where r_1 and r_2 are simultaneously low at some point of time. In other words, the coverage gap can be accurately represented by the following property that considers exactly the above scenarios:

$$U_1: \qquad G(\ (\ \neg r_1 \ \wedge \ \neg r_2\ ) \ \Rightarrow \ X \ g_1\ )$$

We have $R_2 \wedge U_1 \Rightarrow A_3$, and therefore U_1 closes the coverage gap between R_2 and A_3. In general, our aim is to determine the *weakest* set of temporal properties that close the coverage gap between the RTL specification and the architectural intent. This intent is formally expressed below. $\square$

Definition 6.5. [Strong and weak properties:]
A property $\mathcal{F}_1$ is stronger than a property $\mathcal{F}_2$ iff $\mathcal{F}_1 \Rightarrow \mathcal{F}_2$ and $\mathcal{F}_2 \not\Rightarrow \mathcal{F}_1$. We also say that $\mathcal{F}_2$ is weaker than $\mathcal{F}_1$. $\square$

Definition 6.6. [Coverage Hole in RTL Spec:]
A coverage hole in the RTL specification is a property $\mathcal{R}_H$ over $\mathcal{AP}_\mathcal{R}$, such that $(\mathcal{R} \wedge \mathcal{R}_H) \Rightarrow \mathcal{A}$ is valid, and there exists no property, $\mathcal{R}'_H$, over $\mathcal{AP}_\mathcal{R}$ such that $\mathcal{R}'_H$ is weaker than $\mathcal{R}_H$ and $(\mathcal{R} \wedge \mathcal{R}'_H) \Rightarrow \mathcal{A}$ is valid. In other words, we find the weakest property that suffices to close the coverage hole. Adding the weakest property strengthens the RTL specification in a minimal way. $\square$

Since $\mathcal{AP}_\mathcal{A} \subseteq \mathcal{AP}_\mathcal{R}$, each property of the architectural intent is a valid property over $\mathcal{AP}_\mathcal{R}$. The following theorem characterizes the coverage hole.

Theorem 6.7. *The coverage hole in the RTL specification is unique and is given by $\mathcal{A} \vee \neg\mathcal{R}$.*

Proof: Let $\mathcal{R}_H = \mathcal{A} \vee \neg\mathcal{R}$. It is easy to see that $(\mathcal{R} \wedge \mathcal{R}_H) \Rightarrow \mathcal{A}$, and therefore $\mathcal{R}_H$ closes the coverage hole.

Let $\mathcal{R}'_H$ be a property such that $\mathcal{R}'_H$ is weaker than $\mathcal{R}_H$ and $(\mathcal{R} \wedge \mathcal{R}'_H) \Rightarrow \mathcal{A}$. Since $\mathcal{R}'_H \not\Rightarrow \mathcal{R}_H$, there exists a run, π, that satisfies $\mathcal{R}'_H$ but not $\mathcal{R}_H$.

Suppose π satisfies $\mathcal{R}$. Then by the definition of $\mathcal{R}'_H$, π satisfies $\mathcal{A}$. But if π satisfies $\mathcal{A}$, then π must satisfy $\mathcal{R}_H$ (by the definition of $\mathcal{R}_H$). This is a contradiction.

Otherwise, suppose π does not satisfy $\mathcal{R}$. Therefore π satisfies $\neg\mathcal{R}$, and again π must satisfy $\mathcal{R}_H$ (by the definition of $\mathcal{R}_H$). Again we have a contradiction. Therefore $\mathcal{R}_H$ is the unique weakest property that closes the coverage gap. $\square$

There is an intuitive explanation of the coverage hole as defined by Theorem 6.7. The goal of the design intent coverage analysis is to find those behaviors that refute $\mathcal{A}$ but satisfy $\mathcal{R}$, that is, those behaviors that satisfy:

$$\varphi = \neg\mathcal{A} \wedge \mathcal{R}$$

The property representing the coverage hole must reject exactly these behaviors, hence the property is $\mathcal{A} \vee \neg\mathcal{R}$ which is $\neg\varphi$. The following example demonstrates the notion of a *coverage hole* in our formulation.

Example 6.8. Let us again consider the arbiter of Example 6.3. We had seen that the coverage gap lies in A_2 and A_3. By Theorem 6.7 we have the coverage hole in the RTL specification as:

$$R_H: \quad (\,(A_2 \wedge A_3) \vee \neg(R_1 \wedge R_2 \wedge R_3)\,)$$

We can also write the same coverage hole as the conjunction of the following two properties:

$$R_H': \quad (\,A_2 \vee \neg(R_1 \wedge R_2 \wedge R_3)\,)$$
$$R_H'': \quad (\,A_3 \vee \neg(R_1 \wedge R_2 \wedge R_3)\,)$$

In other words, we can examine the coverage of each architectural property separately and produce a set of properties representing the coverage hole. $\square$

Typically, the coverage hole, $\mathcal{A} \vee \neg\mathcal{R}$, will contain signals belonging to $\mathcal{AP}_\mathcal{R} - \mathcal{AP}_\mathcal{A}$. To demonstrate the part of the architectural intent that is not covered by the RTL specification, we need a further level of abstraction. The definition of the *uncovered architectural intent* is as follows.

Definition 6.9. [Uncovered architectural intent:]

The uncovered architectural intent is a property $\mathcal{A}_H$ over $\mathcal{AP}_A$, such that $(\mathcal{R} \wedge \mathcal{A}_H) \Rightarrow \mathcal{A}$ is valid, and there exists no property $\mathcal{A}'_H$ over $\mathcal{AP}_A$ such that $\mathcal{A}'_H$ is weaker than $\mathcal{A}_H$ and $(\mathcal{R} \wedge \mathcal{A}'_H) \Rightarrow \mathcal{A}$ is valid. In other words, we find the weakest property over $\mathcal{AP}_A$ that suffices to close the coverage hole. $\square$

6.2.2 How Should We Present the Coverage Hole?

Theorem 6.7 gives us a formalism for computing the coverage hole, but does not present the missing properties in a meaningful way. Our aim is to present the coverage hole and the uncovered architectural intent before the designer in a form that is syntactically close to the architectural intent and is thereby amenable to visual comparison with the architectural intent.

The expressibility of the logic used for specification does not always permit a succinct representation of the coverage hole. In such cases, we prefer to present the coverage hole as a succinct set of properties that closes the coverage gap, but may be marginally stronger than the actual coverage gap. The following example highlights this intent.

Example 6.10. We consider the coverage of A_3 by R_2 in the specifications given in Example 6.3. By Theorem 6.7, the coverage gap between A_3 and R_2 is given by the property:

$$\varphi = A_3 \vee \neg R_2$$

The knowledge that the above property is satisfiable does not convey any meaningful information to the designer. On the other hand, consider the property U_1 of Example 6.4:

$$U_1: \qquad G(\ (\ \neg r_1 \ \wedge \ \neg r_2\ ) \ \Rightarrow \ X \ g_1\ )$$

U_1 is stronger than φ, but is able to represent the coverage gap more effectively than φ. This is because, the designer can visually compare U_1 with A_3 and see what remains to be covered.

It is also important to be able to preserve structural similarity with the architectural intent when we present the coverage hole. For example, the property U_1 can also be written as:

$$G(\ r_1 \ \vee \ r_2 \ \vee X \ g_1\ )$$

or as:

$$G(\ (\ \neg X \ g_1 \ \wedge \ \neg r_1\ ) \ \Rightarrow \ r_2\ )$$

These representations are logically equivalent to U_1, but are not visually similar to A_3. Preserving structural similarity is a very important issue in presenting the gaps between formal property specifications. $\square$

The SpecMatcher tool uses two key algorithms for structuring the uncovered architectural intent. The first algorithm enables the computation of the real time bounded terms in the coverage gap and then *pushes* these terms into the syntactic structure of the architectural properties to obtain the uncovered part. The second algorithm takes architectural properties having unbounded temporal operators (such as G, F and U) and systematically weakens them into structure preserving decompositions. It then checks the components that remain to be covered. The following example illustrates this approach.

Example 6.11. Let us consider the coverage of the property:

$$A_2' : \quad G(\, (\, r_2 \wedge z\,) \,\Rightarrow\, X(\, (\, g_2 \wedge \neg\, g_1\,)\, U\, \neg r_2\,)\,)$$

by the RTL property:

$$R_2' : \quad G(\, (\, r_1 \wedge \neg r_2\,) \,\Leftrightarrow\, X\, g_1\,)$$

and the fairness property $R_1 = GF(\,\neg r_2\,)$. We find that $R_1 \wedge R_2' \Rightarrow A_2'$ is not valid, which establishes that A_2' is not covered. However, we can decompose A_2' into a conjunction of two properties, namely:

$$A_{2a}' : \quad G(\, (\, r_2 \wedge z\,) \,\Rightarrow\, X(\, g_2\, U\, \neg r_2\,)\,)$$
$$A_{2b}' : \quad G(\, (\, r_2 \wedge z\,) \,\Rightarrow\, X(\, \neg\, g_1\, U\, \neg r_2\,)\,)$$

The property A_{2b}' is covered by R_2' and R_1. The coverage hole is therefore in the property A_{2a}' which is a more accurate (weaker) representation than A_2'. □

6.3 The Intent Coverage Algorithm

We are now in a position to present the algorithms for producing a structure preserving form of the coverage gap. The SpecMatcher tool takes each formula $\mathcal{F}_A$ from the architectural intent $\mathcal{A}$ and finds the coverage gap, $\mathcal{G}$, for $\mathcal{F}_A$, with respect to the RTL specification $\mathcal{R}$.

Algorithm 6.1 Find_Coverage_Gap($\mathcal{F}_A$,$\mathcal{R}$)

---

1. Compute $\mathcal{U} = \mathcal{F}_A \vee \neg \mathcal{R}$

2. If $\mathcal{U}$ is not valid then

   a) Unfold $\mathcal{U}$ to create a set of uncovered terms, $\mathcal{U}_M$, that approximates the coverage gap;

b) Use universal quantification to eliminate signals belonging to $\mathcal{AP}_\mathcal{R} - \mathcal{AP}_\mathcal{A}$,

c) $\mathcal{F}_\mathcal{U}$ = Call Push_Term($\mathcal{F}_\mathcal{A}, \mathcal{U}_M, 1$);

d) $\mathcal{G}$ = Call Relax_ArchitecturalIntent($\mathcal{F}_\mathcal{U}$);

3. Return $\mathcal{G}$;

The first step determines the coverage gap formula $\mathcal{U}$ in terms of RTL variables. If $\mathcal{U}$ is valid then $\mathcal{F}_\mathcal{A}$ is covered. Otherwise we need to find an abstraction of $\mathcal{U}$ over the architectural variables that is syntactically close to $\mathcal{F}_\mathcal{A}$. The second step of the algorithm performs this task. This step is further divided into four steps. We shall now describe each of these steps with examples and correctness proofs.

6.3.1 Unfolding the Coverage Gap

In Step 2(a), we recursively unfold $\mathcal{U}$ and generate a disjunction of terms, $\mathcal{U}_M$, that contain only Boolean subformulas and Boolean subformulas guarded by a finite number of X (next) operators. We guarantee that $\mathcal{U}_M$ is as strong as the coverage gap, $\mathcal{U}$. By this approximation we eliminate all unbounded temporal operators from $\mathcal{U}$, which helps us to push the terms in $\mathcal{U}_M$ into the syntactic structure of $\mathcal{F}_\mathcal{A}$. Before we describe the unfolding step we present the definitions of an *X-pushed formula*, *X-guarded formula* and *X-depth of an operator* within a formula.

Definition 6.12. [X-pushed formula:]
A formula is said to be X-pushed if all the X operators in the formula are pushed as far as possible to the left. □

Definition 6.13. [X-guarded formula:]
A formula is said to be X-guarded if the corresponding X-pushed formula starts with an X operator whose scope covers the whole formula. □

Definition 6.14. [X-depth of an operator:]
The X-depth of an operator within a formula is the number of X operators whose scope covers the operator in the X-pushed form. □

Example 6.15. Let us consider the temporal property

$$\mathcal{P} = ((X\, p)\, U\, (X\, X\, q))\, \wedge\, (X\, F\, r)$$

The X-pushed form of $\mathcal{P}$ is:

$$\mathcal{P}_X \;=\; X\,((\,p\,U\,(\,X\,q\,))\,\wedge\,(\,F\,r\,))$$

Now $\mathcal{P}$ is an X-guarded formula because the corresponding X-pushed formula $\mathcal{P}_X$ starts with an X operator whose scope covers the whole formula. $\mathcal{P}$ contains two unbounded temporal operators, U and F. The X-depth of both U and F is 1. $\square$

Our methodology for decomposing $\mathcal{U}$ into a disjunction $\mathcal{U}_M$ of terms is as follows. It is known that any LTL property can be recursively unfolded over time steps to create an equivalent property over Boolean formulas and X-guarded LTL formulas [47]. For example, the property $p\,U\,q$ may be rewritten as:

$$q \vee [\,p \wedge X(\,p\,U\,q\,)\,]$$

after one level of unfolding, and as:

$$q \vee [\,p \wedge X(\,q \vee (\,p \wedge X(\,p\,U\,q\,)\,)\,)\,]$$

after two levels of unfolding. After k-level unfolding, we can distribute the X operators over the Boolean operators and the $\wedge$ operator over the $\vee$ operator to obtain a disjunction of terms, where each term consists of a conjunction of Boolean literals, X-guarded Boolean literals, and X-guarded temporal formulas. For example, $p\,U\,q$ can be rewritten as follows after two levels of unfolding:

$$(\,q\,) \vee (\,p \wedge X\,q\,) \vee (\,p \wedge X\,p \wedge X\,X(\,p\,U\,q\,)\,)$$

Since a temporal formula has a finite number of members in its closure [38], it follows that for every temporal property such a decomposition begins to produce similar X-guarded subformulas after a well defined number of unfolding steps. During the unfolding process we check whether such a fixpoint has been reached.

Once we have the disjunction of the terms, we drop the terms that contain any temporal operator other than X and call the remaining formula as $\mathcal{U}_M$. Dropping terms from the disjunction ensures that $\mathcal{U}_M$ is at least as strong as $\mathcal{U}$. $\mathcal{U}_M$ contains only Booleans and X-guarded Booleans, which is appropriate for Step 2(b) and Step 2(c).

Example 6.16. Consider the property $\mathcal{U} = (p\,U\,q) \vee X\,F(\neg p)$. After one step of unfolding the property looks like:

$$\mathcal{U}^1 \;=\; q \vee (p \wedge X(\,p\,U\,q\,)) \vee X\,F(\neg p)$$

Dropping the present state variables and removing an X from the remaining sub formulas of $\mathcal{U}^1$ generates:

$$\mathcal{U}' = (\ p\ U\ q\ )\ \vee\ F(\neg p)$$

Since $\mathcal{U}'$ is not equivalent to $\mathcal{U}$, we have not yet reached the fixpoint. After the next step of unfolding of $\mathcal{U}'$ and then dropping the present state variables and removing an X from the remaining sub formulas yields:

$$\mathcal{U}'' = (\ p\ U\ q\ )\ \vee\ F(\neg p)$$

which is equivalent to $\mathcal{U}'$. Since any further decomposition will generate the same formula, this is the fixpoint. After two steps of unfolding the property $\mathcal{U}$ becomes:

$$\mathcal{U}^2 = q\ \vee\ (\ p\ \wedge\ X(\ q\ \vee\ p\ \wedge\ X(\ p\ U\ q\ )))\ \vee$$
$$X(\ \neg p\ \vee\ X\ F(\neg p)\ )$$

It can be rewritten in the form of disjunction of the terms as follows:

$$(\ q\ )\ \vee\ (\ p\wedge\ X\ q\ )\ \vee\ (\ p\ \wedge\ X\ p\ \wedge$$
$$X\ X(\ p\ U\ q\ ))\ \vee\ (\ X(\neg p)\ )\ \vee\ (\ X\ X\ F(\neg p)\ )$$

Now dropping the terms containing any temporal operators other than X yields the following set of terms as $\mathcal{U}_M$:

$$\mathcal{U}_M = \{\ q,\ p\ \wedge\ X(q),\ X(\neg p)\ \}$$

It may be noted that $\mathcal{U}_M$ may contain variables belonging to $\mathcal{AP}_\mathcal{R} - \mathcal{AP}_\mathcal{A}$. In order to obtain the uncovered architectural intent, we need to eliminate these variables. This is done in Step 2(b). $\square$

Theorem 6.17. *The property represented by the set of terms $\mathcal{U}_M$ closes the coverage hole for $\mathcal{F}_\mathcal{A}$.*

Proof: Since $\mathcal{U}_M$ is generated from $\mathcal{U}$ by dropping some terms from the disjunctive form, it follows that $\mathcal{U}_M$ is at least as strong as $\mathcal{U}$, and therefore closes the coverage hole. $\square$

6.3.2 Elimination of Non-architectural Signals

In Step 2(b) of the coverage algorithm, we universally eliminate the variables in $\mathcal{AP}_\mathcal{R} - \mathcal{AP}_\mathcal{A}$ from the property $\mathcal{U}_M$. The following theorem establishes that after the abstraction, $\mathcal{U}_M$ still closes the coverage hole.

Theorem 6.18. *The property represented by the set of terms $\mathcal{U}_M$ after universal abstraction closes the coverage hole for $\mathcal{F}_A$.*

Proof: Follows from the fact that universal abstraction results in a property that is at least as strong as the original property. $\square$

6.3.3 The Term Distribution Algorithm

We showed in Section 6.3.1 that $\mathcal{U}_M$ is a disjunction of terms that together close the coverage gap for $\mathcal{F}_A$. Our target is to represent this coverage gap as a set of properties that are structurally similar to $\mathcal{F}_A$. We achieve this objective by distributing the terms in $\mathcal{U}_M$ into the structure of $\mathcal{F}_A$. The following theorem shows that this approach is theoretically sound.

Theorem 6.19. *The property $\mathcal{U}_M \vee \mathcal{F}_A$ is at least as weak as $\mathcal{U}_M$ and closes the coverage gap for $\mathcal{F}_A$.*

Proof: Since $\mathcal{U}_M \Rightarrow \mathcal{U}_M \vee \mathcal{F}_A$ is valid, it follows that $\mathcal{U}_M \vee \mathcal{F}_A$ is at least as weak as $\mathcal{U}_M$. To show that $\mathcal{U}_M \vee \mathcal{F}_A$ closes the coverage gap, we present the following argument.

Since $\mathcal{U}_M$ is as strong as $\mathcal{U}$ (by Theorem 6.18), it follows that $\mathcal{U}_M \vee \mathcal{F}_A$ is as strong as $\mathcal{U} \vee \mathcal{F}_A$. Since $\mathcal{U} = \mathcal{F}_A \vee \neg \mathcal{R}$, it follows that $\mathcal{U} \vee \mathcal{F}_A = \mathcal{U}$. Therefore $\mathcal{U}_M \vee \mathcal{F}_A$ is as strong as $\mathcal{U}$ and closes the coverage gap. $\square$

The remainder of this section presents the methodology for distributing the terms in $\mathcal{U}_M$ into the structure of $\mathcal{F}_A$. The intuitive idea is to push the terms to the sub formulas having similar variables. However, $\mathcal{U}_M$ may also contain some terms that contain variables from $\mathcal{AP}_A$ other than those in $\mathcal{F}_A$. Let us denote these variables by $\mathcal{EV}$ (for *entering variables*).

The Function *Push_Term($\mathcal{F}$, $\mathcal{U}_M$, θ)* pushes the terms in $\mathcal{U}_M$ into the syntactic structure of property, $\mathcal{F}$. To intuitively explain the working of this function, consider a case where $\mathcal{F}$ is of the form $f \Rightarrow g$. Let $Var(f)$ and $Var(g)$ denote the set of variables in f and g respectively. We compute the universal abstraction of $\mathcal{U}_M$ with respect to $Var(g)$ and recursively push the restricted terms (containing only variables in $Var(g)$) to g. We then compute the universal abstraction of $\mathcal{U}_M$ with respect to $Var(f) \cup \mathcal{EV}$ and recursively push the restricted terms to f. The decision to push terms containing entering variables to the left of the implication is heuristic (but correct, since we could push them either way). Pushing the entering variables in the other side may give us another form of the uncovered architectural intent (and we may like to present both forms).

In case $\mathcal{F}$ is of the form $f \wedge g$, we push each term of $\mathcal{U}_M$ to both f and g.

In case $\mathcal{F}$ is of the form $f\ U\ g$ or $F\ f$ or $G\ f$, we maintain the list of variables of $\mathcal{U}_M$ with $\mathcal{F}$ for later use by the *variable weakening algorithm* (Step 2(d)).

The third argument, θ of the function *Push_Term* specifies whether $\mathcal{U}_M$ should be considered in disjunction with $\mathcal{F}$ (in which case $\theta = 1$) or in conjunction with $\mathcal{F}$ (denoted by $\theta = 0$). At the root level, we always have $\theta = 1$ (since we compute $\mathcal{F}_A \vee \mathcal{U}_M$). However the semantics of negation sometimes require us to recursively call *Push_Term()* with $\theta = 0$.

Algorithm 6.2 Push_Term($\mathcal{F}$, $\mathcal{U}_M$, θ)

---

case($\mathcal{F} \equiv (f \Rightarrow g)$) :

    if $(\theta = 1)$ {   Push_Term(f, $\neg UABS(\mathcal{U}_M, Var(f) \cup \mathcal{EV})$, 0);
                                        Push_Term(g, $UABS(\mathcal{U}_M, Var(g))$, 1); }

    if $(\theta = 0)$ {   Push_Term(f, $\neg\mathcal{U}_M$, 1);
                                          Push_Term(g, $\mathcal{U}_M$, 0) ; }

case($\mathcal{F} \equiv (f \vee g)$) :

    if($\theta = 1$) {   Push_Term(f, $UABS(\mathcal{U}_M, Var(f) \cup \mathcal{EV})$, 1);
                                        Push_Term(g, $UABS(\mathcal{U}_M, Var(g))$, 1); }

    if($\theta = 0$) {   Push_Term(f, $\mathcal{U}_M$, 0);
                                        Push_Term(g, $\mathcal{U}_M$, 0); }

case($\mathcal{F} \equiv (f \wedge g)$) : { Push_Term($f$, $\mathcal{U}_M$, θ);
                                              Push_Term(g, $\mathcal{U}_M$, θ); }

case($\mathcal{F} \equiv (\neg f)$) :   Push_Term($f$, $\neg\mathcal{U}_M$, $\neg\theta$);

case($\mathcal{F} \equiv (X\ f)$) : { $\mathcal{U}_{M-x} = XABS(\mathcal{U}_M)$;
                                       Push_Term(f, $\mathcal{U}_{M-x}$, θ); }

case($\mathcal{F} \equiv (G\ f)$ or $(F\ f)$ or $(f\ U\ g)$) :
                      Maintain the list of variables of $\mathcal{U}_M$ with $\mathcal{F}$ for later
                      use by variable weakening algorithm (Step 2(d));

case($\mathcal{F} \equiv p \in \mathcal{AP}_T$) :
  if($\theta = 1$)   Replace p by $p \vee \mathcal{U}_M$;
  if($\theta = 0$)   Replace p by $p \wedge \mathcal{U}_M$;

The functions $UABS()$ and $XABS()$ are as follows:

UABS(φ, $\mathcal{SV}$):   This function takes a set of terms, φ, and a set of variables $\mathcal{SV}$ as input and universally eliminates the set of variables given by $\mathcal{AP}_A - \mathcal{SV}$ from φ. It returns the property given by the union of the abstracted set of terms.

XABS(φ):   This function takes a set of terms, φ, extracts those terms that are within the scope of one or more X operators, and returns these terms after dropping the most significant X operator.

Lemma 6.20. *The property $\mathcal{F}_\mathcal{U}$ produced by Push_Term($\mathcal{F}$, $\mathcal{M}$, θ) is as strong as $\mathcal{F} \vee \mathcal{M}$ when $\theta = 1$ and as weak as $\mathcal{F} \wedge \mathcal{M}$ when $\theta = 0$.*

Proof: We construct the proof by induction on the subformulas of $\mathcal{F}$. In the base case we have an atomic proposition (say p). In this case, we return $p \vee \mathcal{M}$ when $\theta = 1$ and $p \wedge \mathcal{M}$ when $\theta = 0$, and thereby satisfy the lemma.

Suppose $\mathcal{F}$ is of the form $f \Rightarrow g$. When $\theta = 1$, we replace f by the property Push_Term(f, $\neg UABS(M, Var(f) \cup \mathcal{EV})$, 0) and we replace g by Push_Term(g, $UABS(M, Var(g))$, 1).

By induction hypothesis, Push_Term(f, $\neg UABS(M, Var(f) \cup \mathcal{EV})$, 0) is as weak as $f \wedge \neg UABS(M, Var(f) \cup \mathcal{EV})$.

Likewise, we have that Push_Term(g, $UABS(M, Var(g))$, 1) is as strong as $g \vee UABS(M, Var(g))$.

Hence it follows that Push_Term($f \Rightarrow g$, M, 1) is as strong as:

$$(f \wedge \neg UABS(M, Var(f) \cup \mathcal{EV}) \Rightarrow g \vee UABS(M, Var(g)))$$

which is the same as:

$$(\neg f \vee UABS(M, Var(f) \cup \mathcal{EV}) \vee g \vee UABS(M, Var(g)))$$

Now by the definition of $UABS$, we have:

$$UABS(M, Var(f) \cup \mathcal{EV}) \vee UABS(M, Var(g))$$

is as strong as M. Hence Push_Term($f \Rightarrow g$, M, 1) is as strong as $(f \Rightarrow g) \vee M$ as required by the lemma.

Let us now consider the case when $\mathcal{F}$ is of the same form, and $\theta = 0$. Now, Push_Term($f \Rightarrow g, M, 0$) is recursively computed as:

$$Push\_Term(f, \neg M, 1) \;\Rightarrow\; Push\_Term(g, M, 0)$$

By induction hypothesis, Push_Term($f, \neg M, 1$) is as strong as $f \vee \neg M$ and Push_Term($g, M, 0$) is as weak as $g \wedge M$. Hence Push_Term($f \Rightarrow g, M, 0$) is as weak as $(f \vee \neg M) \Rightarrow (g \wedge M)$, that is $(f \Rightarrow g) \wedge M$, as required by the lemma.

The proof follows a similar style when $\mathcal{F}$ is of the forms $f \vee g$, $f \wedge g$ or $\neg f$. When $\mathcal{F}$ is of the form $X\ f$, the proof directly follows from the definition of $XABS$ function and the distribution rule in Push_Term.

For the remaining cases, namely when $\mathcal{F}$ is of the forms $G\ f$, $F\ f$ or $f\ U\ g$, the algorithm simply returns $\mathcal{F}$. Since $\mathcal{F}$ is as strong as $\mathcal{F} \vee \mathcal{U}_M$ and as weak as $\mathcal{F} \wedge \mathcal{U}_M$, the lemma (vacuously) holds for these cases as well. $\square$

Theorem 6.21. *Push_Term*$(\mathcal{F}_A, \mathcal{U}_M, 1)$ *returns a property* $\mathcal{F}_\mathcal{U}$ *that closes the coverage hole.*

Proof: From Lemma 6.20, we have that Push_Term$(\mathcal{F}_A, \mathcal{U}_M, 1)$ is as strong as $\mathcal{F}_A \vee \mathcal{U}_M$. Along with Theorem 6.19, we have that $\mathcal{F}_\mathcal{U}$ closes the coverage hole. $\square$

Example 6.22. Let us consider the architectural property A and the RTL property R as given below, where $r_1, r_2, g_1, g_2 \in \mathcal{AP}_A$.

$$A: \quad r_1 \Rightarrow X(\ g_1\ U\ g_2\ )$$
$$R: \quad (\ r_1 \wedge r_2\ ) \Rightarrow X(\ g_1\ U\ g_2\ )$$

After step 2(b) of algorithm 6.1 we have the following set of terms for $\mathcal{U}_M$:

$$\mathcal{U}_M = \{\neg r_1,\ X(g_2),\ r_1 \wedge r_2 \wedge \neg X(g_2)\}$$

The distribution of the above terms into the parse tree of A by Algorithm 6.2 is shown in the Figure 6.7.

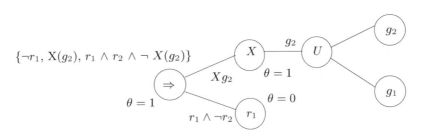

Fig. 6.7. Working of Algorithm Push_Term

After the execution of the Push_Term algorithm $\mathcal{F}_\mathcal{U}$ is represented in the following form:

$$\mathcal{F}_\mathcal{U}: (\ r_1 \wedge \neg r_2\ ) \Rightarrow X(\ g_1\ U\ g_2\ )$$

We also have the sole element g_2 in the variable list corresponding to the node U. This list is to be used in Step 2(d) of the intent coverage algorithm. $\square$

It is interesting to note that pushing the terms abstracted from $\mathcal{F}_A \vee \neg\mathcal{R}$ into the structure of $\mathcal{F}_A$ yields a better representation of the coverage gap as compared to pushing the terms abstracted from $\neg\mathcal{R}$ into the structure of $\mathcal{F}_A$, though both approximate $\mathcal{F}_A \vee \neg\mathcal{R}$.

Example 6.23. Let us revisit the previous example. Suppose we take $\neg R$ instead of $A \vee \neg R$ to generate the terms. We get the following as $\mathcal{U}_M$:

$$\mathcal{U}_M : \{r_1 \wedge r_2 \wedge \neg X(g_2)\}$$

Here, after the Push_Term algorithm we have the property A as $\mathcal{F}_\mathcal{U}$ and an empty variable list at the U node because $UABS$ fails to abstract out terms from $\mathcal{U}_M$ due to absence of the extra terms generated from A, in $\mathcal{U}_M$. $\square$

6.3.4 Weakening Unbounded Temporal Properties

Algorithm Push_Term affects only those portions of the architectural intent that are time bounded. Since the *bounded until* operator of real time LTL can be expressed using the X and Boolean operators, the fragment of LTL that consists of Boolean operators and all bounded temporal operators can be appropriately handled by Push_Term.

For properties having the unbounded temporal operators, namely G (always), F (eventually), and U (until), we use heuristics to decompose the property into weaker fragments and then return those fragments that are not covered by the RTL specification. Within the weaker fragments, we may again use Algorithm Push_Term to compute the uncovered architectural intent more accurately.

Our intuitive idea in this step is to systematically weaken the intermediate uncovered architectural intent, $\mathcal{F}_\mathcal{U}$, while ensuring that it still closes the coverage hole. We use two kinds of structure preserving weakening methods in our tool. Both the methods are based on the observation that a property can be weakened by appropriately weakening or strengthening a variable occurrence. The task of determining whether a weakening or strengthening is required that results in weakening of the property can be performed by a simple examination of the parse tree of the property[2]. We provide some details of the two methods used in our tool for weakening.

Weakening by substitution: A variable instance in its native form can be weakened by substituting it with *true* and strengthened by substituting

---

[2] Note that we weaken/strengthen only one instance of a variable at a time, and not for all instances of that variable in the property. In case the instance of the variable being weakened/strengthened is vacuous, the resulting property will be as weak as the original.

it with *false*. The effect of such a substitution on the property which contains the instance depends on the polarity of the variable instance in that property.

We can systematically weaken or strengthen a property by substituting variable instances with true/false appropriately. The approach is as follows. After choosing a variable (that is under unbounded temporal operators) and performing the weakening substitution, we examine whether the weakened property still closes the coverage hole. If so, then we recursively attempt to weaken it further.

Substitution does not disturb the syntactic structure of the remaining property, and hence the uncovered architectural intent produced in the end is visually comparable to the original architectural intent. For example, consider the following property:

$$\varphi: \qquad G(\ a \Rightarrow X(\ c\ U\ d\ )\ )$$

Suppose we want to weaken the property φ, by substituting the variable instance c with true/false. Here, in order to weaken φ, we need to weaken the variable instance c and since the variable instance is in positive polarity in φ, hence it should be substituted by *true*. After the substitution we get the following weakened property φ':

$$\varphi': \qquad G(\ a \Rightarrow X\ F\ d\ )$$

Variable Weakening: In addition to substitutions, a property can also be weakened by augmenting it with new literals, that is, by replacing a variable instance with the conjunction or disjunction of the variable with other literals.

The choice of new literals to be introduced into a property can be given by the user, or may be chosen by examining the terms originating from the unbounded temporal subformulas. In the latter, we use the variable list corresponding to each variable occurrence (obtained during Push_Term) for weakening of that variable occurrence. We disjunct or conjunct a literal with a variable occurrence depending on whether weakening or strengthening of the variable occurrence weakens the formula.

We demonstrate the idea with the following property:

$$\varphi: \qquad G(\ (\ a\ U\ b\ ) \Rightarrow (\ c\ U\ d\ )\ )$$

Suppose we want to weaken the property by augmenting a new literal $\neg e$ with the variable instance a. Here we have to strengthen the variable instance a for weakening of φ since the polarity of the variable is negative in φ. So we need to replace the variable instance with the conjunction of the variable and the new literal. The resulting weakened property will be as follows:

$$\varphi': \qquad G(\, (\, (\, a \wedge \neg e \,)\, U\, b\,)\, \Rightarrow (\, c\, U\, d\,)\,)$$

6.3.5 Special Treatment of Invariant Properties

Properties of the form $\mathcal{F}_A = G(\varphi)$ (also called invariants) are very common in formal property specifications, hence we treat the coverage of such formulas separately. The following algorithm sketch outlines our approach.

Algorithm 6.3 Coverage of Invariants

---

1. Let ψ denote the property formed by conjuncting the collection of invariant properties in $\mathcal{R}$. ψ is computed by syntactic parsing of the RTL specification.

2. We let $\mathcal{U} = G(\,\varphi\,) \vee \neg G(\,\psi\,)$ and compute $\mathcal{U}_M$ as in Step 2(a) of Algorithm 6.1.

3. These terms are then pushed into φ using Push_Term to obtain the weakened property, φ'.

4. We return $G(\varphi')$ as the intermediate uncovered architectural intent and apply Step 2(d) of Algorithm 6.1 on the subformulas of $G(\varphi')$ to further refine the uncovered architectural intent.

The following theorem shows that the uncovered architectural intent obtained in this way closes the coverage hole.

Theorem 6.24. *The property, $G\,\mathcal{F}_{\mathcal{U}}$, returned by Algorithm 6.3 closes the coverage gap of $G(\varphi)$ with respect to $\mathcal{R}$.*

Proof: From Lemma 6.20, $\mathcal{F}_{\mathcal{U}}$ is as strong as $\varphi \vee \mathcal{U}_M$. Also by Theorem 6.18, $\mathcal{U}_M$ is as strong as $\mathcal{U}$ and hence $G\,\mathcal{F}_{\mathcal{U}}$ is as strong as $\mathcal{F}_1 = G(\,\varphi \vee \mathcal{U}\,)$. Now from Theorem 6.7, we have:

$$\mathcal{U} = G\,\varphi \vee \neg G\,\psi = G\,\varphi \vee F\,\neg\psi$$

Since $G\,\psi$ is as weak as $\mathcal{R}$ (because we considered only the invariant terms), it follows that $\mathcal{U}$ is as strong as the coverage hole, and thereby closes the coverage hole. To prove the result, we shall show that the property returned by Algorithm 6.3 is as strong as $\mathcal{U}$ and is able to close the coverage hole as well. Substituting this property for $\mathcal{U}$ in $\mathcal{F}_1$ we get:

$$\mathcal{F}_1 = G(\ \varphi \vee G\ \varphi \vee F\ \neg\psi\ ) = G(\ \varphi \vee F\ \neg\psi\ )$$

Now in every run π that satisfies $\mathcal{F}_1$, either φ is true at all states, or there exists some state that satisfies $\neg\psi$. In the first case π satisfies $G\ \varphi$, and in the second case π satisfies $F\ \neg\psi$. In both cases, π satisfies $\mathcal{U}$. Hence $\mathcal{F}_1$ is as strong as $\mathcal{U}$ and therefore is able to close the coverage hole. $\square$

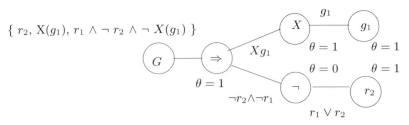

Fig. 6.8. Term distribution

Example 6.25. Let us return to the specifications shown in Example 6.3 and let us consider the coverage of A_3 by R_2:

$$A_3: \quad G(\ (\ \neg r_2\ )\ \Rightarrow\ X\ g_1\ )$$
$$R_2: \quad G(\ (\ r_1 \wedge \neg r_2\ )\ \Rightarrow\ X\ g_1\ )$$

Since both properties are invariants, we use Algorithm 6.3. In the first step, we compute ψ as $(\ r_1 \wedge \neg r_2\ )\ \Rightarrow\ X\ g_1$ The set of terms, $\mathcal{U}_M$ obtained by unfolding $A_3 \vee \neg G\psi$ are as follows:

$$\mathcal{U}_M = \{r_2,\ X(g_1),\ r_1 \wedge \neg r_2 \wedge \neg X(g_1)\ \}$$

We now call Push_Term to distribute the terms in $\mathcal{U}_M$ into the parse tree of A_3 past the G operator (as shown in Fig 6.8). The figure shows how the terms in $\mathcal{U}_M$ distribute across the different operators in A_3. The value of θ shown besides the nodes indicate whether the incoming terms are in conjunction or disjunction with the property rooted at that node. The uncovered part of A_3 after applying Push_Term is given by:

$$G(\ (\ \neg(r_1 \vee r_2)\ )\ \Rightarrow\ X\ g_1\ )$$

Let us now revisit the coverage problem of Example 5, where we examine the coverage of A_2' by R_2' and the fairness property $R_1 = GF(\ \neg r_2\ )$:

$$A_2': \quad G(\ (\ r_2 \wedge z\ )\ \Rightarrow\ X(\ (\ g_2 \wedge \neg g_1\ )\ U\ \neg r_2\ )\ )$$
$$R_2': \quad G(\ (\ r_1 \wedge \neg r_2\ )\ \Leftrightarrow\ X\ g_1\ )$$

In this case also, all properties are invariants, hence we apply Algorithm 6.3. However, in this case Algorithm 6.3 fails to weaken A_2' and returns A_2'. Therefore, we systematically weaken A_2' and search for weaker properties that still close the coverage gap. In this case, we find that substituting 0 for g_1 in A_2' gives us the property:

$$A_{2a}' : \qquad G(\ (\ r_2 \wedge z\ )\ \Rightarrow\ X(\ g_2\ U\ \neg r_2\ )\ )$$

which is weaker than A_2', but still closes the coverage gap. By our approach, no further structure preserving weakening of A_{2a}' can close the coverage gap, hence we report A_{2a}' as the uncovered part of A_2'. $\square$

6.4 Soundness of the Intent Coverage Algorithm

Theorem 6.26. *The coverage gap $\mathcal{G}$ returned by Find_Coverage_gap($\mathcal{F_A}, \mathcal{R}$) closes the coverage hole.*

Proof: There are two steps in Algorithm 6.1. Step 1 generates the exact coverage hole $\mathcal{U}$ that closes the coverage hole by Theorem 6.7.

From Theorem 6.17 and Theorem 6.18, we have that after Step 2(b), the set of terms in $\mathcal{U}_M$ closes the coverage hole. By Theorem 6.21, the intermediate uncovered architectural intent property after step 2(c), $\mathcal{F}_\mathcal{U}$, still closes the coverage hole. In step 2(d), we perform weakening of $\mathcal{F}_\mathcal{U}$ but maintaining the property that it remains at least as strong as the coverage hole in $\mathcal{R}$ with respect to $\mathcal{F_A}$. This concludes the proof. $\square$

6.5 Multi-property Representation of the Coverage Gap

We have observed that in some cases it is possible to represent the coverage gap as a collection of two or more properties that are individually similar to the architectural property, and are able to close the coverage gap taken together. The following example demonstrates one such case.

Example 6.27. Let us consider the architectural property $\mathcal{F_A}$ and the RTL specification $\mathcal{R}$, as given below:

$$\mathcal{F_A}: \quad r_1 \Rightarrow X(\ g_1 \wedge g_2\ )$$
$$\mathcal{R}: \quad (\ r_1 \wedge r_2\ ) \Rightarrow X\ g_1$$

By Algorithm 6.2, the terms of the coverage gap, $\mathcal{F_A} \vee \neg\mathcal{R}$, to be pushed to the left of the implication in $\mathcal{F_A}$ is $\neg r_1$ and the terms to be pushed to the right of the implication in $\mathcal{F}_A$ is $X\ g_1 \wedge X\ g_2$. This does not change $\mathcal{F}_A$. Relaxing $\mathcal{F}_A$ using our heuristics also does not give any better result.

Now, consider the following two properties:

$$A_1': \quad (\ r_1 \wedge \neg r_2\ ) \Rightarrow X\ g_1$$
$$A_2': \quad r_1 \Rightarrow X\ g_2$$

The conjunction of these two properties closes the coverage gap and also both are weaker than the architectural property $\mathcal{F}_A$. Clearly, these two properties taken together is a better representation of the coverage gap. $\square$

We now present a heuristic methodology for decomposing the coverage gap as a set of properties. For this purpose we compute $\mathcal{U} = \neg \mathcal{R}$ in Step 2(a) of the coverage algorithm (instead of $\mathcal{U} = \mathcal{F}_A \vee \neg \mathcal{R}$). In Step 2(c) we use a modified version of Push_Term which uses a different approach only for the cases where $\mathcal{F}$ is of the form $f \Rightarrow g$ or $f \vee g$. Recall that in the original version we used universal abstraction to determine the set of terms to be pushed towards the subformula f and set of terms to be pushed towards g respectively. Universal abstraction guarantees that no new terms are added and therefore the modified property is guaranteed to be as strong as $\mathcal{F}_A \vee \neg \mathcal{R}$ and closes the coverage gap.

In the modified version of Push_Term we treat the cases $f \Rightarrow g$ or $f \vee g$ as follows. We perform an *existential* abstraction on $\mathcal{U}_M$ to eliminate the variables not belonging to $Var(\mathrm{f}) \cup \mathcal{EV}$ and push the resulting terms towards f to obtain a property $\mathcal{F}_{\mathcal{U}_L}$. Likewise, we perform an existential abstraction on $\mathcal{U}_M$ to eliminate the variables not belonging to $Var(g)$ and push the resulting terms towards g to obtain a property $\mathcal{F}_{\mathcal{U}_R}$. We found that in many cases, in spite of the fact that $\mathcal{F}_{\mathcal{U}_L}$ and $\mathcal{F}_{\mathcal{U}_R}$ are individually weaker than $\mathcal{F}_A \vee \neg \mathcal{R}$, the property $\mathcal{F}_{\mathcal{U}_L} \wedge \mathcal{F}_{\mathcal{U}_R}$ is as strong as $\mathcal{F}_A \vee \neg \mathcal{R}$ and therefore closes the coverage gap. However, since this is not always true, we attempt this heuristic approach when the original methodology fails to represent the coverage gap with a property that is weaker than $\mathcal{F}_A$ and apply the result after checking that $\mathcal{F}_{\mathcal{U}_L} \wedge \mathcal{F}_{\mathcal{U}_R} \Rightarrow \mathcal{F}_A \vee \neg \mathcal{R}$ is valid, that is, $\mathcal{F}_{\mathcal{U}_L} \wedge \mathcal{F}_{\mathcal{U}_R}$ is as strong as the coverage gap.

Example 6.28. Let us return to the previous example. The term resulting from $\mathcal{U}' = \neg \mathcal{R}$ is $r_1 \wedge r_2 \wedge \neg X(g_1)$. Now applying the modified version of Push_Term, we get $r_1 \wedge r_2$ to be pushed in the left hand side of the implication, resulting in a formula:

$$\mathcal{F}_{\mathcal{U}_L}: \quad (\ r_1 \wedge \neg r_2\ ) \Rightarrow X(\ g_1 \wedge g_2\ )$$

Similarly, $\neg X(g_1)$ is pushed to the right side of the implication, resulting in:

$$\mathcal{F}_{\mathcal{U}_R} : \qquad r_1 \Rightarrow X(g_2)$$

We can see that both $\mathcal{F}_{\mathcal{U}_L}$ and $\mathcal{F}_{\mathcal{U}_R}$ are individually weaker than $\mathcal{F}_{\mathcal{A}}$ but their conjunction is stronger than $\mathcal{U}$ and hence closes the coverage gap between $\mathcal{F}_{\mathcal{A}}$ and $\mathcal{R}$. □

6.6 SpecMatcher – The Intent Coverage Tool

SpecMatcher is our in-house tool for design intent coverage over LTL specifications. The architecture of the tool is shown in Fig 6.9.

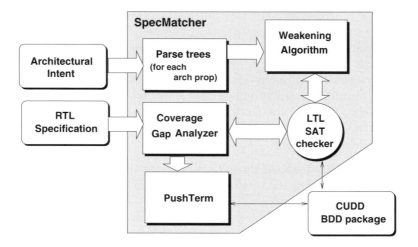

Fig. 6.9. The Architecture of SpecMatcher

The tool accepts two specifications, namely the architectural intent and the RTL specification. Both specifications should be in LTL. It produces the uncovered architectural intent (in LTL) as the output.

SpecMatcher uses an in-house LTL satisfiability checker for checking the primary coverage question. It treats the LTL SAT checker as an oracle – therefore we can easily integrate it with more advanced SAT tools. Our in-house LTL SAT checker is BDD-based and uses the CUDD BDD package to maintain the BDDs.

The terms of the coverage gap are symbolically maintained as BDDs. The PushTerm algorithm uses the BDD representation of the terms.

The weakening heuristics also use the LTL SAT checker to determine whether the weakened specification still closes the coverage gap.

SpecMatcher does not provide a push-button solution to the intent coverage problem. The user is provided with options for choosing the heuristics for weakening and substitution, or for using the multi-property coverage heuristics. These heuristics present the same coverage gap in different forms. We believe that in practice validation engineers should study the different representations and use the one that seems most readable in the context of the specific design.

We used SpecMatcher on several test cases from Intel. The results are presented in our papers (see the bibliographic notes).

6.7 Priority Cache Access – *A Closer Look*

We will now consider a more detailed version of our introductory example, namely that of the priority cache access block. Fig 6.10 shows the architecture of the cache access logic. The specification is as follows:

- There are four request inputs, $r1, \ldots, r4$, for four independent requesting modules.

- Each of these four modules also assert write-enable signals $w1, \ldots, w4$ respectively.

- The signals $h1, \ldots, h4$ are inputs from the cache that indicate whether the page requested by $r1, \ldots, r4$ respectively are present in the cache.

- When the output signal *hold* is asserted the arbitration logic stops accepting any request.

- The signals $d1, \ldots, d4$ indicate whether the page requested by $r1, \ldots, r4$ respectively is ready.

The architectural intent requires that $r1$ and $r2$ have higher priority than $r3$ and $r4$. This may be translated into the following architectural property:

If two devices with different priorities make requests for the cache control unit with the higher priority device making the request before the lower priority one, the device with higher priority will always have its page ready at the output, before the device with lower priority.

This architectural level requirement can be expressed by four properties for the four different ways in which a high priority request can come with a low priority one. For example, when we consider $r1$ (high priority) with $r3$ (low priority) we have the following property:

A_1: $G(((r1 \ \wedge \ \neg r2 \ \wedge \ \neg r3) \ \wedge \ X(r1 \ U \ r3)) \ \Rightarrow \ (\neg d3 \ U \ d1))$

We have the following fairness conditions on the inputs and outputs, which require the requesting device to hold the request line high until the page becomes ready:

Q_1: $G(\ r1 \Rightarrow (\ r1\ U\ d1\ )\ )$
Q_2: $G(\ r3 \Rightarrow (\ r3\ U\ d3\ )\ )$

We are also given the following two fairness restrictions:

Q_3: $G(\ \neg r1 \Rightarrow (\ \neg d1 \wedge X(\ \neg d1\ )\ )\ )$
Q_4: $G(\ \neg r3 \Rightarrow (\ \neg d3 \wedge X(\ \neg d3\ )\ )\ )$

These restrictions state that whenever a device is idle (that is, not requesting), its page ready signal will be low in that cycle and its next cycle.

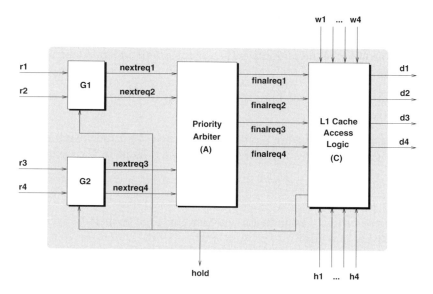

Fig. 6.10. Memory arbitration with L1 cache

Figure 6.10 shows the architecture of this logic in terms of four modules. G1 and G2 are round-robin arbiters. The arbiters process requests in a round robin fashion provided that the *hold* signal is not asserted. Whenever *hold* is asserted they ignore their request lines. The variable *lastreq* represents the state of G1 indicating which device was granted in the last round. Below we present the RTL properties for G1. The properties for G2 are similar.

R_{G1_1}: $G((r1 \wedge \neg r2 \wedge \neg hold) \Rightarrow (nextreq1 \wedge X(\neg lastreq)))$
R_{G1_2}: $G((\neg r1 \wedge r2 \wedge \neg hold) \Rightarrow (nextreq2 \wedge X(lastreq)))$

R_{G1_3}: $G((r1 \land r2 \land \neg hold \land lastreq)$
$$\Rightarrow (nextreq1 \land X(\neg lastreq)))$$
R_{G1_4}: $G((r1 \land r2 \land \neg hold \land \neg lastreq)$
$$\Rightarrow (nextreq2 \land X(lastreq)))$$
R_{G1_5}: $G(hold \Rightarrow (\neg nextreq1 \land \neg nextreq2))$
R_{G1_6}: $mutex(\ nextreq1,\ nextreq2\ )$

The priority arbiter in Figure 6.10 selects requests in the priority order, namely G1-highest and G2-lowest. The specification for this module is given below:

R_{A_1}: $G(\ nextreq1 \Rightarrow X\ finalreq1\ )$
R_{A_2}: $G(\ nextreq2 \Rightarrow X\ finalreq2\ )$
R_{A_3}: $G((\neg nextreq1 \land \neg nextreq2 \land nextreq3) \Rightarrow X\ finalreq3)$
R_{A_4}: $G((\neg nextreq1 \land \neg nextreq2 \land nextreq4) \Rightarrow X\ finalreq4)$
R_{A_5}: $mutex(\ finalreq1,\ \ldots\ ,\ finalreq4\ )$

The cache access logic in Figure 6.10 directly interacts with the cache and works as follows. Whenever a cache miss occurrs in a read transfer, the cache logic keeps the corresponding request pending in a wait buffer. *bfull* is an internal signal which notifies the wait buffer full condition. It sends the *hold* signal to the G1 and G2 arbiters whenever buffer full condition is high.

The following three properties are required in the RTL specification of the cache access logic to repond correctly to device 1. We have similar properties of the cache access logic to specify the reponses to the other requesting devices.

R_{C_1}: $G((finalreq1 \land h1) \Rightarrow X\ d1)$
R_{C_2}: $G((finalreq1 \land w1 \land \neg h1) \Rightarrow X\ d1)$
R_{C_3}: $G((finalreq1 \land \neg w1 \land \neg h1 \land \neg bfull) \Rightarrow X\ F\ d1)$

For the buffer full condition, the following two properties are required.

R_{C_4}: $G(\ bfull \Leftrightarrow hold\ )$
R_{C_5}: $G(\ bfull \Rightarrow X\ F(\ \neg bfull\ )\ )$

Finally we have the mutual exclusion property for the ready signals.

R_{C_6}: $mutex(\ d1,\ d2,\ d3,\ d4\ )$

If we enumerate all the RTL properties noted above, then we have a collection of 36 RTL properties that constitute the RTL specification. It may be noted that we do not yet have the RTL implementation for the modules G1, G2, A and C. The objectives of the intent coverage problem are (a) to check whether these properties cover the architectural properties such as A_1, and (b) to determine the coverage gap if it exists.

The answer to the primary coverage question for the architectural property A_1 is negative, that is, A_1 is not implied by the RTL specification. In the scenario when the higher priority request is a read operation that results in a cache miss, A_1 is not guaranteed by the RTL properties. A counter example scenario that demonstrates this gap is as follows:

> *Device 1 has requested for a page read at time t, while device 2 and device 3 are idle. At time $t + 1$, device 3 makes a request and device 1's request results in a cache miss. Note that the* hold *signal is low in both the cycles.*

> *Therefore at time $t + 2$, device 1 does not get the corresponding page ready signal but device 3's request results in a cache hit and hence at time $t + 4$ device 3 has its page ready signal available at the output.*

We analyzed the specs with SpecMatcher. The tool enables us to find the following coverage gap property (that is, the uncovered architectural intent):

$$A_H:\ G(\ (\ (\ r1\ \wedge\ \neg r2 \wedge\ \neg r3\ )$$
$$\wedge\ (\ (\ X(\ \neg w1\ )\ \wedge\ X(\ \neg h1\ )\ )\ \vee\ hold\ )$$
$$\wedge\ X(\ r1\ U\ r3\ )\ )$$
$$\Rightarrow\ (\ \neg d3\ U\ d1\ )\ )$$

The property represents the set of scenarios that are not covered by the RTL specification and is visually very similar to the property A_1.

6.8 Concluding Remarks

If we study the growth in the size of digital circuits over the last decade, it is quite astonishing that chips have been successfully taped out with so few bugs. *How do we do it?*

The answer lies in *methodology* – the use of a very systematic approach towards design and making the best use of existing CAD tools.

At the heart of this methodology is the designer's ability to systematically and recursively partition the design functionality into the functionality of the component modules. This helps in coding the smaller blocks and verifying them locally before intergrating them into the main design.

Creating the test environment for a design requires the same creativity – the ability to partition the functionality of the test bench into the functionalities of the components that constitute the environment of the design-under-test. Not surprisingly, most recent test bench languages and verification methodologies (such as RVM of Synopsys and VMM of ARM/Synopsys) emphasize the need of a structured and hierarchical approach for designing test benches.

Today, do we use a similar hierarchical approach for developing a formal verification test plan? Possibly not. Most practitioners of FPV treat FPV as a one-step approach – develop the formal specification and ask a FPV tool to verify it over the given design. This approach limits the use of FPV to the unit level beyond which it is currently infeasible.

Design intent coverage helps in enabling the notion of *specification refinement*. It helps the validation engineer in developing a feasible formal verification test plan. The validation engineer can start with the high level (architectural) specification and recursively decompose the specification into sets of smaller properties over the components of the design. At each step, the intent coverage tool enables her to verify whether (a) the decomposition is correct, and (b) whether the decomposition is complete, that is, whether it has sufficient properties to cover the original intent.

This is the missing formal validation flow that must run in parallel with the design flow – that of using the design decomposition to perform the specification decomposition. This flow only uses the block structure of the implementation and is thereby scalable. The hierarchical decomposition terminates at the unit level, where it is easy to design *and easy to verify.*

Design intent coverage is not a push-button solution. It relies on the validation engineer's ability to use the hierarchy of the design to partition the specification and develop the properties at the lower levels of the design hierarchy. Intent coverage only exposes the gaps and suggests the missing properties. It is up to the validation engineer to interpret the gaps and use the information to strengthen the formal verification plan.

6.9 Bibliographic Notes

Several early literatures on compositional verification has studied the decomposition of specifications into properties that describe the behavior of small parts of the system [90, 64]. It is possible to show that if the system satisfies each local property, and if the conjunction of the local properties implies the overall specification, then the complete system must satisfy this specification as well.

The notion of analyzing the coverage gap between temporal specifications and the design intent coverage paradigm was developed through a collaboration between our research group and Intel. The basic idea was first presented in [14]. The algorithms for computing the coverage gap properties were outlined in [44] and presented in details in [15]. Several additions to the basic design intent coverage paradigm have been made recently [16, 46].

Test Generation Games

There are several important issues in the frontier between simulation and FPV. Assuming that the future of design validation lies in the symbiotic co-existence of these two technologies, it is important to investigate how each may benefit the other. Our intention is to present some insights into this area, but before we proceed to do so, we must study the recent developments in simulation-based test environments.

The industry appears to be moving towards simulation frameworks that employ a coverage driven random test generation approach, as opposed to the traditional approach of writing directed tests. The traditional approach, which is still prevalent in many chip design companies, is as follows:

1. The validation engineers identify the set of interesting behaviors where they believe bugs may hide. The golden functional requirements in such cases is sometimes refered to as the *specification*, and the interesting scenarios are called *coverage points*.

2. A test plan is developed to cover the set of interesting behaviors. The test plan consists of a set of tests that together achieve the desired level of coverage. The test plan is a document (say, in English).

3. Directed tests are written for each item of the test plan. For each test, the validation engineer has to write a test bench to mimic the environment of the design-under-test (DUT) and drive the appropriate stimuli to reach the desired coverage points. In a complex design, the task of finding out the sequence of stimulus that reaches a coverage point is often non-trivial and requires careful planning and understanding of the design.

4. Simulation is performed for each of the directed tests.

For example, let us consider the task of verifying a master interface in the ARM AMBA AHB protocol. The Bus architecture is shown in Fig 7.1. The

Bus supports several different types of transfers – read and write transfers, single and burst transfers, incremental bursts and bounded bursts, locked transfers. It also supports split transfers and retry responses. It supports both types of endianness. Since the master can initiate and participate in these different types of transfers, a test plan for the master must cover many different cases. In order to drive meaningful inputs into the module, a test environment for the master must correctly mimic the behavior of the other components such as the slave devices, the Bus arbiter, the bridge to the APB Bus, and other contending master devices. The validation engineer must be aware of the protocol followed by the other components while developing a test for a given master interface, which is a non-trivial task.

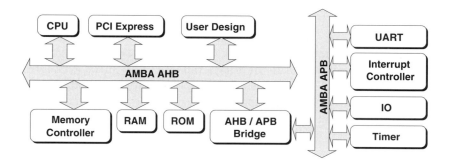

Fig. 7.1. An Arm AMBA Architecture

As the design complexity increases, we face two problems, namely (a) the number of directed tests required to cover all the interesting behaviors grow alarmingly, and (b) the task of writing each directed test becomes more complex, thereby requiring more time from the validation engineer.

7.1 Constrained Random Test Generation

As the design complexity keeps pace with Moore's Law, the traditional approach of writing directed tests is loosing its relevance – it either requires too much time, or it fails to achieve a meaningful level of coverage. This has paved the way for randomized test environments.

If we use an unconstrained random test generation approach, then it becomes hard to guarantee that the test bench will drive valid input sequences. For example, to verify a AMBA AHB master interface the test bench must mimic a AHB compliant environment – one that drives valid AHB compliant inputs during the course of a transaction. It is hard to guarantee this if we randomly choose the values of the inputs in each cycle.

Therefore it is prudent to use randomization at the real *choice points* of the environment. For example, there are several choice points for the test environment of a AMBA AHB slave interface. We may randomly choose the transfer type (read/write, single/burst, etc) and the master which initiates the transfer. Once we have chosen the transfer type and the master, the test bench initiates the transfer by imitating the transfer request from the master in a protocol compliant manner. The length of the transfer, the values of the address and data lines are the choice points during the transfer. None of these choices are unconstrained. For example, to choose the given slave in a memory mapped system, the values of the address lines have to be constrained so that the address lies in the map of the device.

Monitoring the coverage is a very important component of the constrained random verification approach. Since the tests are generated randomly, we must know which behaviors have been covered. Recent test bench languages such as the test bench language of SystemVerilog enables a validation engineer to specify the coverage goal, it allows her to bias the random choice at choice points by associating weights with the choices, and it allows her to monitor the status of the coverage.

7.1.1 Layered Verification Methodologies

In the last few years, several EDA companies are promoting a structured hierarchical methodology for developing the simulation test environment for complex digital designs. These include the *e-Reuse Methodology* (eRM) of Verisity, the *Reference Verification Methodology* (RVM) of Synopsys, and *Verification Methodology* (VMM) of ARM and Synopsys. The focus of these methodologies is in enabling the validation engineer to develop a set of functional models for the components of a large design and then using these models in a constrained random test generation environment.

Recent design and verification languages such as SystemVerilog and *e* support the style of verification suggested by these methodologies. SystemVerilog, for example, is a combination of a design language (Verilog) and a test bench language. The test bench language supports a large set of powerful constructs that facilitate the development of behavioral models for the components of a large design. For example, the SystemVerilog test bench language supports fork-join constructs, synchronization constructs such as semaphores, communication constructs such as channels, and a fully object oriented design style.

How do the behavioral models help in building a randomized test environment? The test bench language supports features that enable the validation engineer to call a randomizer to choose the values of signals under specified constraints. The models of the components use this feature at each choice point. The test bench instantiates those models that constitute the environment of the DUT and handles the high-level choices (such as the choice of the transaction type) randomly. The models randomly generate their responses in a protocol compliant manner and this guarantees that the inputs driven into the DUT are also protocol compliant.

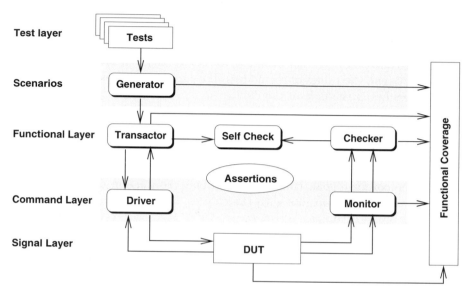

Fig. 7.2. A Layered Test Architecture

These ideas have been integrated into a systematic layered approach for creating a test environment. The typical layers in a RVM or VM compliant test architecture are shown in Fig 7.2. We briefly outline the meaning of these layers in the context of a Bus protocol (say, AMBA AHB).

- *Signal Layer.* The sensitization of the DUT takes place in this layer. The interfaces of the DUT with the test bench and the monitor are defined at this layer.

- *Command Layer.* This layer drives the unit level transactions through the DUT (using the signal layer). For example, typical transfer types in AMBA AHB includes, *idle, read, write, burst-read,* and *burst-write.* The command layer needs to be aware of the Bus protocol, and must drive the transactions in a protocol compliant manner. This task becomes significantly easier when the models of the components of the architecture are used. For example, when we need to drive a write transfer to a slave DUT, it makes our task significantly easier if we have the behavioral model of a protocol compliant master to drive the transfer.

- *Functional Layer.* This layer creates sequences of transactions, and then requests the command layer to drive each transaction in the appropriate sequence. Examples include, *read followed by burst-write*, and *burst-write followed by idle*.

- *Scenario Layer.* Architectural scenarios are generated at this layer. For example, consider the handling of an interrupt. An interrupt can happen in many ways, and can be handled in many ways – each of these cases is a potential test scenario. Other examples include DMA, cache flush, etc. Each scenario consists of a sequence of transactions, which is handled by the functional layer.

- *Test Layer.* The test cases are specified at the highest level of abstraction in this layer. The contents of this layer is mostly application specific.

There is a notable difference between the communication in the signal layer and that in the other layers. In the signal layer, we are actually sensitizing the DUT through the interface. This is a real communication between the DUT and its environment, which needs to be carried down into the hardware. On the other hand, the communication between the higher layers model commands being passed down from a higer layer to a lower layer, or coverage information being passed from the layer into the coverage monitor. These communication channels help us in creating a layered test environment – the channels themselves will not be present in the hardware.

Recent test bench languages such as SystemVerilog provide a communication primitive called *channels* for facilitating the communication between the layers of the test bench. The validation engineer is expected to model all communications between the layers using channels.

Where do the assertions fit into this architecture? Indeed, assertions are needed everywhere. At the command layer, we need assertions to verify whether the transactions are compliant with the Bus protocol. At the functional layer, we need assertions to verify whether the pipelining between successive transfers are being done correctly. At the higher layers, we need assertions to verify whether in each scenario, the application specific properties relevant to that scenario are being satisfied. In Fig 7.2, the assertions are shown between the command layer and the functional layer because the assertion IP for the AMBA protocol (which is application independent) resides in these layers. Application specific assertions can off course be specified and used at the higher layers of the architecture.

7.1.2 The Benefits

Writing the behavioral models for the components of a large design (and validating them) requires a significant quantum of time. However, once the models are complete, we are in an advantageous position. In order to verify a given DUT, we can quickly construct a test bench by instantiating the models that constitute the environment of the DUT. After this, we begin the simulation and no user intervention is required. Since the test bench generates tests randomly, successive iterations of the test bench automatically sensitizes different scenarios and automatically reaches different coverage points. As a result, the test bench covers a large number of coverage points very quickly.

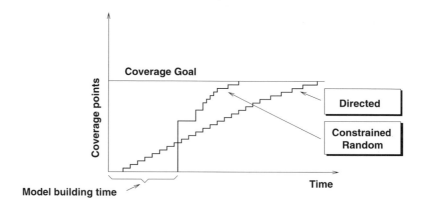

Fig. 7.3. Coverage: Directed versus Randomized

Fig 7.3 shows a typical comparison of the coverage times between the traditional approach of writing directed tests and the new approach of constrained random test generation.

In the traditional approach the coverage of behaviors starts right from the beginning, since the simulation can begin as soon as the validation engineer has developed a test. The slope of the curve depends on the manpower used for writing the tests.

In the constrained randomized approach, simulation starts only after the behavioral models for the components of the environment has been completed. Therefore, there is an initial period when no coverage is achieved – this is the model building phase. The length of this phase depends on the manpower used for developing the models. Once the models are complete and the simulation starts, we see a rapid growth in the coverage. The slope of this curve is dictated by the speed of the simulation and the effectiveness of the randomization in reaching the different coverage points. Significantly, *the slope of this curve is independent of the manpower used*.

Typically the new approach *overtakes* the traditional approach in large designs in terms of speed of coverage. However during the inception of this approach in companies, validation teams are usually under tremendous pressure in the model building phase since they do not have any coverage to show.

7.1.3 The Limitations

What is the probability of a specific scenario being generated randomly? A simple analysis will show that the probability is the product of the probabilities of making the right choices at the choice points that lead to the given scenario. Therefore, if a corner case can be reached through a very special sequence of choices, and the probability of picking the right choice is low in each step, then the probability of coverage of the corner case may become quite low.

Let us consider an example. Let us say that a master device makes a Bus request 10% of the time (that is, with a probability of 0.1). The Bus protocol exhibits some special behavior when none of the masters are requesting – it parks the grant on a default master. It also exhibits some special behavior when *all* the masters are requesting – it sends a hold signal to the bridge connecting to the low performance Bus. We want to cover both scenarios in a system of 5 masters. *What are the probabilities of reaching these two coverage points randomly?*

Our choice points here are the requesting status of the masters.

1. For the first case, the probability that a master is not requesting is 0.9. Hence the probability that none of the masters are requesting is $(0.9)^5$, which is about 0.6. With this probability, it is quite likely that this corner case will be covered.

2. For the second case, the probability that a master is requesting is 0.1, and hence the probability that all of them are requesting is 0.00001, which is very small. With this probability, it is much less likely that this corner case will be covered, unless we run the simulation for a long time.

This is a standard problem in randomized testing – we reach a lot of coverage points very fast, but it takes a lot of time to cover the corner cases that have low probability of occurance. In other words the simulation tends to get trapped in the common behaviors and fails to sensitize some of the interesting uncommon behaviors.

How can we get around this problem? The most common approach is to write directed tests for the special corner cases that were not covered by randomized testing. While this seems to be the most logical thing to do, it does take a significant amount of validation time, because the remaining scenarios are difficult to conceive and finding the right tests to adequately cover these cases is not straight forward. Recent research is attempting to find automated solutions to the problem – the next section touches this topic.

7.1.4 Dynamic Coverage Driven Verification

The intuitive idea being pursued by several researchers is to use the coverage feedback dynamically to find the gap between the coverage goal (that is, what needs to be covered) and the coverage status (that is, what has been covered) and to *automatically guide the test generator towards the gaps*. The idea seems to be applicable at all layers of the test architecture, and at various levels of abstraction.

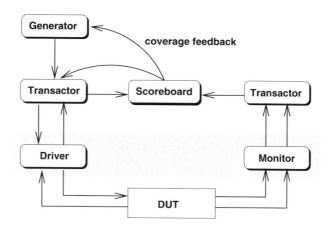

Fig. 7.4. Scoreboard directed coverage

Recent constrained random test frameworks provide two key features to implement the above idea. These are:

1. *The notion of a scoreboard.* A scoreboard keeps track of the coverage points that have been hit during the simulation.

2. *Dynamic modification of weights.* It is possible to associate weights with the different choices at each choice point. The randomzier uses these weights to bias its selection. The weights (and therefore the selection probability) can be modified dynamically.

The above features enable the framework shown in Fig 7.4. A scoreboard is maintained for the interesting coverage points. The feedback from the scoreboard is used to re-distribute the weights on the choices at the random choice points. The key challenge here is to decide how to compute the new weights such that the remaining coverage points become more probable.

7.2 Assertions Viewed as Coverage Points!

FPV cannot verify all the properties that we write. In fact, given the existing state of affairs, FPV can only verify properties at the unit level, where the size of the DUT is small. In the last chapter we presented the notion of *design intent coverage*, which helps the validation engineer in her attempts to formulate a FPV test plan that covers the system level (architectural level) properties of a large design by verifying small local properties on the components of the DUT.

Design intent coverage is an attractive option, and works well in *many* cases, but not in *all* cases. Sometimes we have system level properties for which it is not easy to find a covering set of component properties.

Dynamic assertion based verification is rapidly becoming popular because it fills this void. It helps us to monitor the properties over the simulation run. It flags the violations (and matches) and thereby helps us to debug the design. The methodology does not have serious capacity limitations since we run the assertion checkers over a simulation platform.

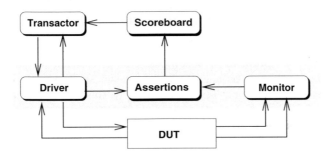

Fig. 7.5. Assertion Coverage

The main limitation of dynamic property verification is that the coverage of the corner cases for which a property was written depends entirely on the test generators ability to reach those corner cases by driving the appropriate inputs into the DUT. If none of these corner cases are reached during simulation, then the property remains *unutilized*, since we have not checked whether the DUT satisfies the property in those corner cases.

In other words, an assertion is a concise specification of a set of corner case behaviors. Therefore, we need to treat assertions as coverage points and make sure that the simulation reaches those behaviors for which an assertion has been written.

One option for solving this problem is by writing directed tests to cover the behaviors for which the property was written. There are at least two major criticisms of this approach:

1. The property encodes the behaviors for which it was written. Having to write the directed tests again to cover the same behaviors is (theoretically) repitition of information, which ideally (that is, given the right technology), should not be necessary.

2. A property can be *triggered* in complex ways. It is not easy for a validation engineer to visualize all types of scenarios in which the property may become relevant.

Our goal is to find an automated solution to this problem. Given an assertion, we want to develop an automatic methodology for guiding the simulation into those behaviors for which the assertion is relevant. The architecture of the desired platform is shown in Fig 7.5. The main challenge is – *How can we automatically guide the simulation to cover the assertion?* This is the focus of this chapter.

7.3 Games with the Environment

Fundamentally, the problem of automatic test generation for covering assertions may be formulated as a game between the DUT and its environment (that is, the test generator). This is because of the fact that the test generator has no control over the behavior of the DUT. In other words, the DUT is a black-box – we can drive inputs into it, but we can neither predict, nor influence its behavior in any way.

Given an assertion, we can partition all possible behaviors of the DUT into two sets – the set of behaviors for which the assertion is relevant and the set of behaviors for which the assertion is not relevant. The assertion is vacuously satisfied in the second set of behaviors. Therefore, validation is meaningful with respect to the given assertion only if the simulation traces a run from the first set.

The objective of the test generator in each cycle of the test generation game is to determine the next set of inputs such that the simulation run traces a behavior from the first set. Since the behavior of the DUT is unknown, we assume that it is adversarial, that is, the DUT attempts to produce outputs such that the simulation traces a behavior from the second set, and thereby satisfies the property vacuously. The test generation algorithm should assume that the DUT is adversarial in this sense. If on the other hand, the DUT actually behaves in a friendly way, then we will reach the coverage goal faster.

So far we have not formalized the notion of *relevance* of a run with respect to a given assertion. We have two broad types of relevance, as explained in the following subsections.

7.3.1 Vacuity Games

Let us consider the following properties for a 2-way round-robin arbiter having request lines, r_1 and r_2, and grant lines, g_1 and g_2:

1. If none of the request lines are high, then in the next cycle both grant lines should be low. We can write this property in SVA as:

   ```
   property P1;
      @ (posedge clk)
      (!r1 && !r2) |- > ##1 !g1 && !g2 ;
   endproperty
   ```

2. If both grant lines are low, and both request lines are high, then the arbiter gives the grant to g_1 in the next cycle. This may be written as:

   ```
   property P2;
      @ (posedge clk)
      (r1 && r2 && !g1 && !g2) |- > ##1 g1 ;
   endproperty
   ```

3. If r_2 is high at time t then either g_2 is asserted at time $t + 1$, or g_2 is asserted in $t + 2$ provided that r_2 is kept high at $t + 1$. We may write this as:

   ```
   property P3;
      @ (posedge clk)
      r2 |- > ##1 (g2 or (!g2 && !r2)
                   or ((!g2 && r2) ##1 g2)) ;
   endproperty
   ```

We explain our notion of vacuity through the above properties.

- The first property, P1, is relevant in only those scenarios where both request lines are low. In all other scenarios (that is, when r_1 or r_2 is high), P1 is satsified vacuously because *the DUT plays no part in satisfying the property in this way*.

  In order to cover the property, P1, the test generator must drive both request lines to low.

- Let us consider the second property, P2. Obviously P2 is satisfied vacuously in all scenarios where r_1 or r_2 is low.

  Is it vacuously satisfied when both r_1 and r_2 are high, but g_1 or g_2 is high? In this case, the antecedent of the implication is false, but the property is satisfied non-vacuously. This is because the DUT played a role in falsifying the antecedent.

  The DUT can satisfy an implication property by satisfying the consequent or by refuting the antecedent. In both cases the property is satisfied non-vacuously, since the behavior of the DUT is responsible for the satisfaction.

  In order to cover the property, P2, the test generator must drive both request lines to high.

- The third property, P3, is vacuously satisfied at all times where r_2 is low. It is satisfied non-vacuously if r_2 is high and the DUT asserts g_2 in the next cycle.

  What if the DUT does not assert g_2 in the next cycle? We have two cases, namely (a) r_2 is de-asserted in the next cycle, and (b) r_2 remains asserted in the next cycle. In the first case, the property is again satisfied vacuously. In the second case, the property is satisfied (non-vacuously) if g_2 is asserted in the subsequent cycle.

  What is the role of the test generator in this case? In order to cover P3, it must drive the request line r_2 to high and study the response of the DUT in the next cycle. If the DUT asserts g_2, then the test generator has a hit. Otherwise, the DUT must drive r_2 to high again, which is now guaranteed to result in a non-vacuous match or fail.

The last case suggests that the test generator needs to be interactive. In other words, the coverage of a temporal property may span over multiple cycles – in each cycle, the test generator needs to study the response of DUT in order to determine the non-vacuous inputs for the next cycle.

Let us now study the game between the DUT and its environment for covering a given property, P. The execution of the `initial`-block of the DUT is the first move of the DUT. If the initial state is sufficient to satify P, then we have a hit. Otherwise, there must exist some assignment to the inputs in the first cycle, such that P is not satisfied. The test generator must choose one such assignment. We now simulate the DUT with that input and study the response of the DUT (that is, the values of the non-inputs) in the next cycle. If P is now satisfied, then we have a hit, otherwise we repeat the process of test generation. At no point of time should the test generator generate an assignment to the inputs for which the property is satisfied regardless of the next response from the DUT.

7.3.2 Realizability Games

In Chapter 4 we presented the notions of *realizability* and *receptiveness*. Both of these concepts are important in the context of test generation.

Unreceptive specifications may create a debugging problem in dynamic property verification. The reason (as outlined in Chapter 4) is as follows. With unreceptive specifications, the failure of a property may take place several cycles after the sensitization of a fault in the DUT. Worse still, a fault may get masked before it is detected unless proper inputs are driven by the test bench. In other words, the fault can be propagated through the cycles by an intelligent choice of inputs until it refutes a property.

The reader is strongly urged to revisit Example 4.1 to refresh the notion of receptiveness and the problems associated with unreceptive specifications.

As demonstrated by Example 4.1, when a fault occurs (that is, the DUT makes a wrong move), the specification may become unrealizable but not unsatisfiable. In other words, the property has not yet been refuted, but with intelligent test generation we are guaranteed to drive the DUT to a state where the property is refuted. This is a very important requirement, since our end goal is to expose bugs.

Given an unrealizable specification, the goal of the test generator is to drive the DUT to a state where the specification is refuted. This may happen in a sequence of steps (with alternate moves between the DUT and the test generator). In each step, the test generator needs to make sure that the inputs generated by it does not make the specification realizable. This game between the DUT and its environment will be called a *realizability* game.

When the DUT produces a fault that renders the specification unrealizable (but not unsatisfiable), then we start a *realizability game* to drive the DUT to a refuting state.

Why dont we simply report a failure when we find that the specification has become unrealizable? The problem is in presentation of the bug. It is hard to demonstrate succinctly why a specification is unrealizable. All counter-example scenarios with respect to the unrealizable specification may not actually be present in the DUT. On the other hand, if we play the realizability game then we can demonstrate a real counter-example after reaching the refutation.

7.4 Intelligent Test Generation for Property Coverage

We now formalize the approach for intelligent test generation[1] when our goal is to cover a given set of properties (that is, to cover the behaviors that are relevant to these properties). We will present the methodology over LTL specifications for ease of presentation.

Formally, we define a module $\mathcal{J}$ as an RTL design block having a set of inputs $\mathcal{I}$, a set of outputs $\mathcal{O}$, an initial block $Init$, and a RTL description $\mathcal{B}$. The execution of the initial block produces the values of the output variables at the beginning of the simulation.

Before we describe the formal procedure of intelligent test generation, we define an *X-pushed formula* and an *X-guarded formula*.

Definition 7.1. [X-pushed formula:]
A formula is said to be X-pushed if all the X operators in the formula are pushed as far as possible to the left. □

Definition 7.2. [X-guarded formula:]
A formula is said to be X-guarded if the corresponding X-pushed formula starts with an X operator whose scope covers the whole formula. □

Example 7.3. Let us consider the temporal property

$$\mathcal{P} \;=\; (\,(\,X\,p\,)\,U\,(\,X\,X\,q\,)\,)\,\wedge\,(\,X\,F\,r\,)$$

The X-pushed form of $\mathcal{P}$ is:

$$\mathcal{P}_X \;=\; X\,(\,(\,p\,U\,(\,X\,q\,)\,)\,\wedge\,(\,F\,r\,)\,)$$

Now $\mathcal{P}$ is a X-guarded formula because the corresponding X-pushed formula $\mathcal{P}_X$ starts with a X operator whose scope covers the whole formula. □

7.4.1 Dynamic Monitoring

The task of monitoring the truth of a given LTL property along a simulation run works as follows (see Section 3.3 for details). If we are required to check a LTL property, φ, from a given time step, t, we rewrite the LTL property as a set of propositions over the signal values at time t and a set of X-guarded LTL properties over the run starting from time $t + 1$. The rewriting rules are standard and are reproduced from Section 3.3:

----

[1] This notion of intelligent test generation was first presented in [13].

$$F\varphi \quad = \quad \varphi \vee XF\varphi$$
$$G\varphi \quad = \quad \varphi \wedge XG\varphi$$
$$p \ U \ q \quad = \quad q \vee (p \wedge X(p \ U \ q))$$

The property checker reads the signal values from the simulation at time t and substitutes these values on the rewritten properties and derives a new temporal property that must hold on the run starting from time $t + 1$, by dropping the leftmost X operator from each X-guarded term.

For example, to check the property $p \ U \ (q \ U \ r)$ at time t, we rewrite it as:

$$(r \vee (q \wedge X(q \ U \ r))) \vee (p \wedge X(p \ U \ (q \ U \ r))$$

If the simulation at time t gives $p = 0$, $q = 1$, $r = 0$, then by substituting these values, we obtain the property $X(q \ U \ r)$. Therefore at time $t + 1$ we need to check the property $q \ U \ r$. We repeat the same methodology on $q \ U \ r$ at time $t + 1$.

7.4.2 Online Test Generation

For automatic test generation, we may choose the values of the input signals at each time step t while monitoring the property. The following definition is useful for characterizing our goal.

Definition 7.4. [Vacuous input vector]
An input vector, $\widehat{I}$, is vacuous at a given state on a run with respect to a property, φ, if φ becomes true at that state on input $\widehat{I}$ regardless of the values of the remaining variables. □

There are two goals of intelligent test generation:

1. To avoid the generation of vacuous input vectors, since these vectors do not trigger the scenarios for which the property was created.

2. To verify whether the property ψ derived from the property φ by substituting the values of signals at time t is unrealizable. If so, the test generator must drive tests to reach a refutation.

It may be noted that a non-vacuous input vector with respect to a conjunction of a set of properties $\mathcal{Q}$ may vacuously satisfy one or more members of $\mathcal{Q}$. The basic idea behind our approach is to treat each property separately, as individual coverage points and generate tests targetting these coverage points.

The validation engineer may define her coverage points as a single property or a combination of two or more properties. Our intelligent test generator should generate tests targetting each coverage point in succession.

7.4.3 Test Generation Algorithms

We will first present the test generation algorithms as a standalone procedure. In the next section we will show how the test generation algorithms can be embedded into a layered verification framework such as RVM or VM.

Procedure SimulateMain outlines our test generation algorithm for a module J and a target property $\mathcal{L}$. It calls procedure GenRefute when the module makes a wrong move and calls GenStimulus in other cycles to produce non-vacuous input vectors. In other words, the procedure GenRefute is called when the specification becomes unrealizable in the presence of a fault – its role is to return input vectors by playing the realizability game. Note that $\mathcal{L}$ can either be a single property or a conjunction of one or more user-defined properties depending on the coverage goal.

Algorithm 7.1 SimulateMain(module: J, property: $\mathcal{L}$)

---

begin

Step 1: Set $\widehat{O}$ = the output vector obtained after execution
        of the initial block $Init$ of J

Step 2: While (not end of simulation) begin

    2.1: Rewrite $\mathcal{L}$ in terms of present state Boolean
        propositions and X-guarded temporal properties

    2.2: Substitute the values of the output variables from
        $\widehat{O}$ in the non X-guarded terms of $\mathcal{L}$ to obtain $\widehat{L}$

    2.3: If $\widehat{L}$ = TRUE, return with success

    2.4: If $\widehat{L}$ = FALSE, return with failure

    2.5: If $\widehat{L}$ is unrealizable, then
        $\widehat{I}$ = GenRefute($\widehat{L}$)
    else
        $\widehat{I}$ = GenStimulus($\widehat{L}$)

    2.6: Obtain $\mathcal{L}'$ from $\widehat{L}$ by substituting $\widehat{I}$ in the non
        X-guarded terms of $\widehat{L}$ and dropping the left-
        most X from each X-guarded temporal property

    2.7: Simulate J with $\widehat{I}$

    2.8: Set $\widehat{O}$ = the output vector obtained after simulation

    2.9: Set $\mathcal{L} = \mathcal{L}'$

    endWhile

end

EndAlgorithm

Algorithm 7.2 Input_Vector GenStimulus(property: $\mathcal{L}$)

// $\mathcal{L}$ is a property over $\mathcal{I}$ and X-guarded terms over $\mathcal{I} \bigcup \mathcal{O}$
begin
Step 1: Rewrite $\mathcal{L}$ as a conjunction of clauses, where each clause
       is a disjunction of Boolean formulas and X-guarded terms
Step 2: Set $\mathcal{P}$ = the Boolean formula obtained from $\mathcal{L}$
       after dropping the X-guarded terms
Step 3: If $\mathcal{P}$ is satisfiable
        Set $\widehat{I}$ = a random input vector that refutes $\mathcal{P}$
   else
        Set $\widehat{I}$ = any random input vector
Step 4: return $\widehat{I}$
end
EndAlgorithm

Algorithm 7.3 Input_Vector GenRefute(property: $\mathcal{L}$)

---

begin
Step 1: Set $\widehat{I}$ = a random input vector
Step 2: For each input variable $p \in \mathcal{I}$
       2.1 Substitute p = 0 in the non X-guarded terms of $\mathcal{L}$
       2.2 If $\mathcal{L}$ remains unrealizable,
          Set the pth bit of $\widehat{I}$ to 0;
     else
          Set the pth bit of $\widehat{I}$ to 1;
    endFor
Step 3: return $\widehat{I}$
end
EndAlgorithm

We explain the working of the algorithm over the case of Example 4.1.

Example 7.5. Recall the arbiter specification of Example 4.1. Initially, g_d is high, while g_1 and g_2 are low. Substituting the values of g_d, g_1 and g_2, the specification remains realizable but does not evaluate to true or false. Let us assume P1 is taken as the first coverage point. We rewrite P1 as:

$$(\neg r_1 \ \vee \ X g_1) \ \wedge \ XG(r_1 \ \Rightarrow \ X g_1)$$

and call GenStimulus with this as the argument. GenStimulus returns a non-vacuous input vector, say $r_1 = 1$, $r_2 = 0$.

The DUT (arbiter) is simulated with this input vector. In response, the DUT asserts g_1, and control returns to Step 2 of SimulateMain, and a match of P_1 is found.

Suppose the test generator now targets the next coverage point, namely, P2. Substituting the values of g_1, g_2 and g_d obtained in the previous cycle, the specification remains realizable, but does not evaluate to true or false. GenStimulus is now called with:

$$(\neg r_2 \vee Xg_2 \vee XXg_2 \vee XXXg_2) \wedge XP_2$$

which is the rewriting of P2. This produces the non-vacuous input vector $r_2 = 1$ and $r_1 = 0$ (r_1 cannot be 1 due to assumption A1).

The DUT is simulated with this input vector and it asserts the output g_2. This time when the control returns to Step 2 and we substitute the value of g_2 in the specification, we find that the specification has become unrealizable, since it now contains the unrealizable property:

$$Xg_d \wedge (r_1 \Rightarrow Xg_1)$$

Therefore we call GenRefute. Evidently, for $r_1 = 1$, the specification remains unrealizable, and hence the vector $r_1 = 1$, $r_2 = 0$ is returned.

When the DUT is simulated with this input vector, it asserts g_1, and a refutation occurs on substitution of g_1 due to the violation of P3. $\square$

7.5 The Integrated Verification Flow

We have developed a prototype tool for intelligent test generation within the layered test architecture proposed in RVM and VM. In our tool, the specification is accepted in LTL. The functions GenStimulus and GenRefute are implemented as oracles that can be called by the transactors through the DirectC interface of SystemVerilog. We outline the broad mechanism for this integration in this Section.

- The GenStimulus and GenRefute procedures are implemented in the generator layer.

- The monitor block samples the output values from the DUT at each simulation clock (Step 2.8 of SimulateMain).

- The coverage block takes a property $\mathcal{L}$ and does the following:

    1. It rewrites $\mathcal{L}$ in terms of present state Boolean propositions and X-guarded temporal properties (Step 2.1 of SimulateMain).

2. It takes the current state of the DUT from the monitor block through the associated channel (Step 2.8 of SimulateMain).

3. It substitutes the values of the output variables from $\widehat{O}$ in the non X-guarded terms of $\mathcal{L}$ to obtain $\widehat{L}$ (Step 2.2 of SimulateMain).

4. If $\widehat{L}$ = TRUE, we return with success (Step 2.3 of SimulateMain).

5. If $\widehat{L}$ = FALSE, we return with failure (Step 2.4 of SimulateMain).

6. Inform the realizability of the property to the generator layer through the respective channel (Step 2.5 of SimulateMain).

7. It waits for the current input vector (driven to the DUT by the Driver) at the input channel from the generator.

8. It obtains $\mathcal{L}'$ from $\widehat{L}$ by substituting $\widehat{I}$ in the non X-guarded terms of $\widehat{L}$ and dropping the leftmost X from each X-guarded temporal property (Step 2.6 of SimulateMain).

- The generator block waits at the input channel from the coverage block. When it gets a transaction containing the realizability information of the current property, it does the following:

  1. If the property $\mathcal{L}$ is unrealizable, it provides the Driver an input through GenRefute procedure (Step 2.5, part 1 of SimulateMain).

  2. If the property $\mathcal{L}$ is realizable, it provides the Driver an input through GenStimulus procedure (Step 2.5, part 2 of SimulateMain).

- The Driver waits for the input vector at the input channel from the generator layer. When it gets input vector $\widehat{I}$, it sends it to the coverage block through an output channel.

- It drives the DUT J with $\widehat{I}$ (Step 2.7 of SimulateMain).

We used this methodology on several verification IPs for standard Bus protocols. The test generation algorithms helped us to reach several complex coverage points in significantly less time as compared to a coverage driven constrained random verification approach.

For example, consider the following property for a PLB Master device in the IBM CoreConnect protocol:

The PLB Master should assert ReadBurst signal for the secondary acknowledged burst read transfer in the cycle following the receipt of ReadBurstTerm from the Slave device for the on-going primary burst read transfer.

The formal property (in LTL) is:

ψ: G((((ReadBurst $\wedge$ PAV $\wedge$ S_Ack) $\wedge$ X(ReadBurst $\wedge$ SAV $\wedge$
S_Ack) $\wedge$ XX (ReadBurstTerm)) $\Rightarrow$ XXX(ReadBurst))

It was hard to reach the behaviors covered by this property using the usual constrained random test generation suite. On using the intelligent test generation approach (with vacuity games), we were able to reach these behaviors remarkably faster.

7.6 Concluding Remarks

Several questions may come up while considering the utility of the methodology presented in this chapter.

- Is it practical to go for property coverage using vacuity games right from the beginning of simulation?

- Should we target test generation for one property at a time, or a collection of properties at a time?

- Is it practical to invoke a realizability checker in every simulation step? Will it degrade simulation performance?

We have some of the answers, but we believe that many more issues will come up when intelligent test generation is brought into practice. Here are a few thoughts on the above questions.

We believe that test generation for property coverage should be started at a time when the constrained random test engine appears to be slowing down in terms of covering new scenarios. This means that most of the common scenarios have been covered and only rare corner case scenarios are left out. At this stage of the simulation we should check the scoreboard to identify those properties for which the coverage is low. Only these properties should be targetted by the intelligent test generator. Initial results show that this gives significant coverage gains.

It is important to note that the structure of the test bench is not changed when we use intelligent test generation. Instead of invoking the randomizer at the choice points for the input signals, we call GenStimulus to return a random non-vacuous input vector. The rest of the test bench remains exactly as before.

A realizability check is expensive, particularly when the specification is large. It is therefore not practical to invoke a realizability check on the specification at every step of the simulation. However in practice, when a fault occurs, typically a small subset of the properties become unrealizable, and as a consequence the whole specification becomes unrealizable. A relatively experienced practitioner of FPV can identify the unreceptive parts of the specification and the properties that are related to the unreceptive properties. It is practical to use a realizability checker on such small collections of properties.

We can eliminate the problem with unreceptive specifications altogether by introducing additional properties to make the specification receptive. This is theoretically possible. It remains to be seen whether it is feasible in practice.

7.7 Bibliographic Notes

A broad span of research from early work on algebraic specifications [58] to more recent work such as [100] addresses the problem of relating tests to formal specifications.

One of the main research directions on this topic is on the use of special constructs for characterizing the input space. In [39], Clarke *et. al.* have proposed a methodology which uses a statically-built BDD to represent the entire input constraint logic. Shimizu *et. al.* [97] describe an approach in which the same formal description is used for collecting coverage information and deriving simulation inputs. There have been many efforts based on constraint solving for test generation as reported in [68].

Offline monitoring for test generation is another another notable approach. In [104], Tasiran *et. al.* use the following approach. Simulation steps are translated (offline) to protocol states transitions using a refinement map and then verified against the specification using a model checker. On the specification state space, the model checker collects coverage information, which is used for future test generation. Another approach is proposed in [65], where the authors present a methodology for automatically determining the appropriate vector constraints, based on the analysis of both the design and the property being checked.

ATPG based techniques have been used in [2, 20] for automatic test generation. In these approaches, the formal properties are synthesized as non-deterministic state machines and ATPG is targeted towards faults representing a discrepancy between the implementation and the state machine representing the specification.

In *counterexample guided test generation*, the model checker is used as an oracle to compute the expected outputs and the counterexamples it generates are used as test sequences. The counterexamples contain both inputs and expected values of outputs and so can automatically be converted to test cases [8, 59]. In another work, Callahan, Schneider, and Easterbrook have presented a methodology using the SPIN model checker to generate tests that cover each block in a certain partitioning of the input domain [29].

The layered and reusable test bench development approach has been adopted in several different verification methodologies. These include RVM [95] of Synopsys, VMM [19] of Arm/Synopsys, and eRM [49] of Verisity.

The IBM CoreConnect protocol specs can be found in [70].

A Roadmap for Formal Property Verification

The benefits of FPV has been established quite emphatically in the last decade. Researchers have analyzed several historically significant failures and have shown that the use of FPV could have detected the bug in the design. Recent practitioners of FPV have been able to uncover interesting flaws in the specifications of complex protocols and intricate bugs in live designs.

Yet, the penetration of FPV into the validation flows of most chip design companies is low. *Why?*

There are both technical and logistic reasons. Design managers are often reluctant to use FPV since the demonstration of its value requires an upfront investment (in time and manpower), and there is no established method for analyzing the gain in productivity as a function of this investment. Moreover, trained manpower in FPV is scarce – as a result there is a (false) perception that FPV is difficult to use, and a very real perception that the technology does not scale well.

We believe that the biggest hindering factor in adopting FPV is in understanding how the technology glues with the rest of the validation flow. In other words, the main challenge is in standardizing the creation of verification plans that integrate simulation and FPV. This is the main puzzle that we have to solve in the years to come.

The puzzle has many pieces. Off course the core model checking technology is at the heart of the problem, but (given its theoretical limitations) it is unlikely to solve the whole puzzle. If it could, then we would not need to glue this technology with traditional simulation.

In this book we have studied some of the other pieces of the puzzle. These include:

1. *Specification refinement.* In Chapter 6 we studied the requirement to create a specification refinement flow parallel to the design flow. Just as we refine the design in steps by decomposing its functionality into that of its blocks and adding level specific details, we must refine the formal specification – starting from the architectural specification, down to the block level specifications, and finally down to the specifications of unit level modules. Formal methods of *design intent coverage* (presented in Chapter 6) enable the validation engineer to establish that the refined specification at a level always covers the golden behavior of the higher levels.

2. *FPV coverage.* Formal verification coverage is one of the most important pieces of our puzzle. It helps us to formally establish the coverage of a FPV test plan. The gaps in FPV coverage must be filled by adding new properties or by reaching the missing behaviors through simulation. FPV coverage is a relatively new area, and most validation engineers do not understand how FPV coverage can be related to functional coverage as targeted by a simulation test plan. We studied some of the advances in FPV coverage in Chapter 5.

3. *Property guided simulation.* Formal specification languages enable the validation engineer to express really complex corner case behaviors in a succinct way. The widely acclaimed benefit of FPV is in checking these properties, since it is extremely hard to hit such corner cases in simulation. *What should we do if FPV runs into capacity issues for such properties?* In Chapter 7, we presented some formal methods for using the property itself to guide the simulation into the corner cases.

The core technology behind all of the above three pieces is one of checking (and maintaining) consistency in formal property specifications. Formal methods for consistency checking in formal specifications were presented in Chapter 4.

The objective of this chapter is twofold – (a) to fit these pieces into an intergrated validation flow, and (b) to touch upon some of the missing pieces that will no doubt assume significance in the years to follow.

8.1 Simulation-based Validation Flow

Let us first study the traditional design flow adopted by most chip design companies for large digital designs. Fig 8.1 shows the typical stages of the design flow. There are two broad parts in the validation flow, namely:

1. *Design Intent Verification.* At this stage, we want to decide *what we want to build.* At this stage, the desired functionality of the chip is known, but we do not know how to achieve the desired functionality. At this stage, the

micro-architecture of the chip is decided, and we want to verify whether the architecture will indeed achieve the desired functionality.

2. *Implementation Verification.* Once the architecture of the chip is decided, the actual implementation of the chip begins. This starts with writing the RTL, followed by the use of synthesis tools to develop the gate level netlist, layout and mask.

Simulation is predominantly used in two places in this flow, namely, to verify the architectural model during design intent verification, and to verify the RTL during implementation verification. At the lower levels of the design flow (that is, gate level and below) extensive simulation is not feasible, except in pockets. For example, transistor level simulation is done for cell characterization, but not for larger circuits (usually).

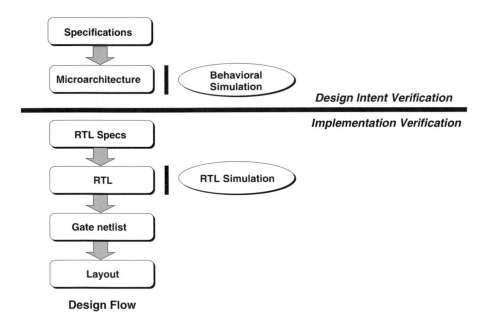

Fig. 8.1. Design and Simulation

The architectural model is typically written at a very high level of abstraction, and in a high level language. Verification at this stage is conceptually challenging, but simulation is not computationally intensive.

The RTL is typically written in a HDL such as Verilog or VHDL. RTL simulation accounts for nearly 70% of the design cycle time of large digital chips. This simulation typically takes place in several stages, namely, unit level validation for individual modules, block level validation for collections of modules, and system level validation for the integrated design. In a large chip design company, the unit level modules may be designed across the globe in different design centers. Design integration and system level verification is typically done in one place.

As an example of the complexities of system level verification, consider the verification of the RTL of a 64-bit processor. Typically, the company will check whether the processor can successfully load an operating system and execute a set of computationally intensive applications *before it tapes out the chip*. During system level verification, the RTL of the processor will be simulated with the complete stack consisting of the operating system and the benchmark applications. Not surprisingly, simulation runs into months.

One of the most glaring gaps in the flow shown in Fig 8.1 is that most chip design companies pass only an English document from the design intent creation phase to the implementation phase. In other words, the RTL designer is typically provided with a specification document, and nothing else.

Misinterpretation of the specification document is one of the major sources of logical bugs in the design. The unit level designer does not believe that a bug exists in her design, since the module matches her interpretation of the specification. After integrating all the modules we find that the design does not match the architectural specification. Debugging shows that there has been a mistake in interpreting the specification, which, more often than not is a result of ambiguous or incomplete specifications. For large chip designs, detecting an architectural violation at this stage of the design flow is really bad news.

Recent design languages are attempting to overcome this problem by enabling the microarchitect and the RTL designer to use the same language. In other words, languages such as System Verilog, SystemC, and Bluespec Verilog attempt to provide abstract modeling capabilities, while retaining all the features of a hardware description language. Several companies also advocate the synthesis of a design directly from an abstract model – it may be quite a while before such frameworks become practical for complex designs.

Fig 8.2 shows the key components in simulation based validation. The test plan is hierarchical in nature – we have unit level tests, block level tests, and system level tests. It may be recalled that in Section 7.1.1 we outlined the notion of layered test bench architectures. Using a structured and layered test bench architecture appears to be the key to design validation.

There is a notable issue here, which will surface when we begin to sketch the formal verification flow.

1. The design is planned top-down. The design is conceived as a collection of communicating blocks. Each block is conceived as a collection of modules.

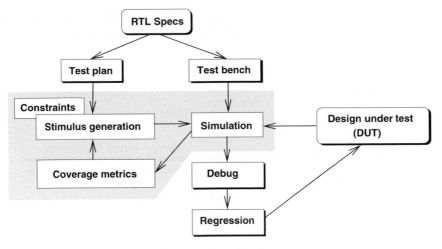

Fig. 8.2. RTL Simulation

This is done recursively, until we have modules of limited functionality (unit modules) that can be coded directly.

2. Verification is done bottom-up. We first validate the unit modules. Then we stitch them into the RTL block and validate the block. In the last step, we validate the whole integrated design.

Therefore, though the architectural functionality is decided first, we are able to verify it at the very end, when all RTL modules have been developed. This is possibly the largest cycle in the design and validation flow.

8.2 Formal Verification Flow

Where in the design flow should we start writing formal properties? Ideally the answer should be – *at the point where we define the design functionality.* Unfortunately, this is not the practice. Today we write properties *only where we can verify them.* This style is largely the outcome of FPV tools having limited capacity.

Fig 8.3 shows the typical stages where FPV may be used. The use of FPV for design intent verification is a new phenomenon. Only recently architects are appreciating the benefits of expressing the design intent formally. Recent property specification languages have facilitated the task of developing a formal architectural specification of a design. Though the notion of formal verification of architectural models is a new phenemenon in chip designs, similar practices have existed in other areas such as automotive control and fly-by-wire systems, where formal specifications of key control laws are verified on abstract models of the machine design before building the machine.

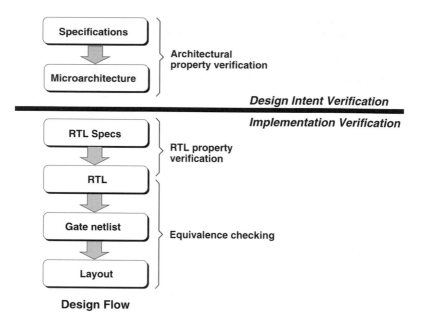

Fig. 8.3. Design and Formal Verification

Most of the existing applications of FPV are on RTL designs. The use of FPV is very limited below the RTL – typically formal equivalence checking is favoured at these levels.

There is a major semantic difference between formal property verification and formal equivalence checking. Formal property verification is a language containment problem. In simple terms, it is possible to have several different implementations of a design – all of which satisfy a given set of properties. These implementations may not be logically equivalent (that is, they may not implement the same Boolean logic), but each of them is an acceptable implementation of the formal property specification. Equivalence checking, on the other hand, requires that the golden model and the given implementation are *logically* equivalent.

The RTL defines the logical functionality of the design. This logic does not change when we synthesize the RTL into a gate level netlist. It also does not change when we translate the gate level netlist into a transistor level netlist. We can therefore use formal equivalence checking at these levels – to verify whether the gate level circuit is equivalent to the RTL, and whether the transistor level circuit is equivalent to the gate level. Today we have an arsenal of formal equivalence checking tools.

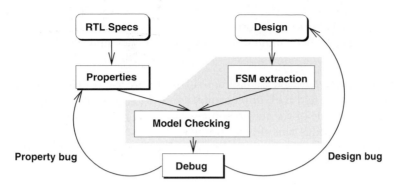

Fig. 8.4. RTL Formal Property Verification

Fig 8.4 shows the typical flow of RTL FPV. The interesting feature in this flow is the existence of the two feedback arcs, one for property bugs and the other for design bugs. Just as the designer may interpret the design specs in an incorrect way, the validation engineer may also interpret the design specs in an incorrect way. As long as both do not interpret the specs in the same incorrect way, FPV will find an inconsistency.

What kinds of bugs do we find using FPV today? We may categorize them as follows.

1. *Bugs found due to schedule advantage.* Typically FPV is used before simulation. Therefore quite a few bugs found by FPV would also have been found during simulation.

2. *Hole in simulation coverage.* Often FPV finds bugs that are not covered by the simulation because the test plan has gaps. Besides exposing the fault, such bugs also enable the validation engineer to strenghten the test plan.

3. *Difficult corner case bugs.* Some bugs are very hard to hit through simulation. FPV can uncover such bugs because it validates the formal specification exhaustively on the design. Discovering such bugs justify the investment made towards building a formal verification test plan.

4. *Performance bugs.* Functionality is not all about logical correctness. Today, performance is a critical component of functionality. It is hard to evaluate worst case performance through simulation, since we need to evaluate all possible behaviors in order to determine the worst case. FPV, by virtue of its exhaustive nature can be used to verify performance parameters naturally.

5. *Impossible bugs.* Sometimes FPV finds bugs at the block level that are not real in the context of the design as a whole. In other words, only some specific inputs to the block sensitizes the bug, and these inputs will never be driven into the block by other blocks in the design. Nevertheless finding such bugs is important, since the block may be reused in some other design where some of the offending inputs may be driven into the block.

In the previous section we had noted that today the bridge between design intent verification and RTL implementation is the specification document, and that this gap is one of the major sources of logical bugs in the design. Formal properties can bridge this divide, provided that we express the architectural intent as formal properties and then carry these properties right down the design validation flow.

8.3 The Three Pillars

We believe that the roadmap for design verification will be around three key technologies (Fig 8.5) – the core FPV (model checking) techniques, the traditional simulation techniques, and specification refinement techniques.

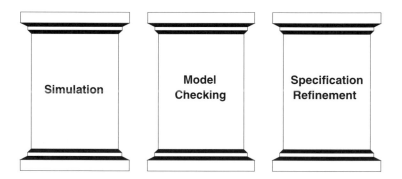

Fig. 8.5. The three key technologies

The important issue here is to figure out the way in which these three technologies will gel to achieve our goal, namely, design validation. Simulation and model checking have been extensively studied over the last couple of decades. Specification refinement is the new methodology that promises to integrate simulation and FPV into a unified validation flow.

The current state of the technology is such that we know what we want to verify, and can also specify the verification requirement formally, but none of the two main technologies (simulation and model checking) enable us to perform this verification adequately. *Why?*

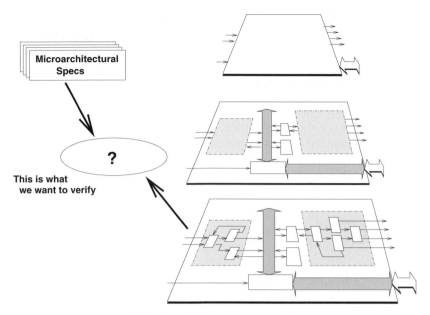

Fig. 8.6. What we want to do

Fig 8.6 shows what we want to do. The notable issues in this flow are as follows.

1. At the beginning of the design cycle, we start with the micro-architectural specification of the design. Today we have the language support to formally express the key features of the specification.

2. The design passes through several levels of abstraction. We never write a single monolithic Verilog code for the whole design. Rather, the design architect conceives a set of abstract blocks interconnected through some simple *glue logic*, that together achieves the functionality of the design. Large functional blocks are similarly decomposed into smaller blocks. This process of design refinement continues until the functionality of each block is simple enough to be coded as a single (unit level) module. In each step of refinement, the designer makes implementation specific choices. The last level, where we have coded all the unit level modules and intergrated them into the design is our RTL implementation.

   This approach has been followed by generations of designers, because it follows the classical notion of divide-and-conquer, which is one of the main methods by which human beings solve complex problems.

The main problem in this approach is that we are not able to verify whether the design refinement steps are correct. Divide and conquer is a sound approach only when the *divide* is done correctly.

3. We want to verify whether the RTL implementation satisfies the microarchitectural specification. Current validation technology falls short of this objective, because the implementation is too big. FPV tools cannot handle this capacity. It is also not possible to achieve reasonable simulation coverage for large designs in feasible time.

Therefore, though we have the language support to write micro-architectural properties, this is not done in practice, since we do not have the tools to verify these properties on the implementation.

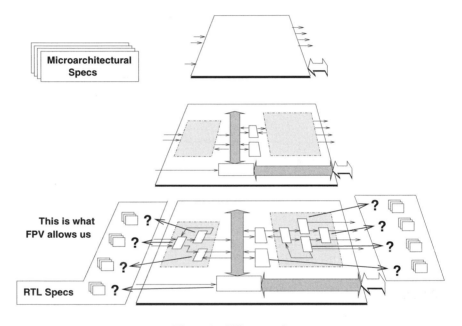

Fig. 8.7. What we do

Fig 8.7 shows what we do as a stop gap solution today. We write formal properties over individual modules and verify them locally using our FPV tools.

Is there any productivity gain in doing this? The answer is yes. When the module size is small, FPV tools can handle them easily and can verify them with almost no user intervention. This is a big advantage, since the task of achieving a similar degree of confidence through simulation can require thousands of test vectors. Moreover, many practitioners of FPV have been able to dig out really complex bugs at this level – bugs that would be hard to hit through simulation.

But this does not solve the main problem. By verifying local properties on individual modules, we have not established that the design as a whole satisfies the microarchitectural specification. This is because we have not established any link between these local properties and the micro-architectural specification.

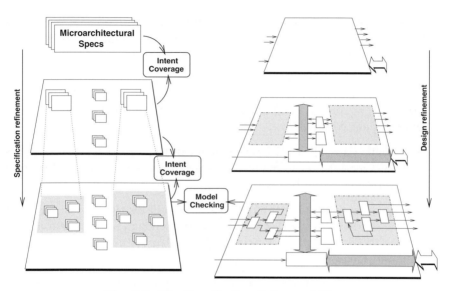

Fig. 8.8. The Specification Refinement Flow

Specification refinement provides this missing link. The basic idea is intuitive and simple. Just as we decompose the design functionality into that of its components, we must decompose the formal microarchitectural specification into the formal specification of the components.

Is this always possible? Theoretically, yes. If we can achieve the functionality of a module, $\mathcal{M}$, through a set of component modules, $M_1, \ldots, M_k$, then it should be possible to theoretically establish that the formal functionality of $M_1, \ldots, M_k$ (manifested through the formal specifications for $M_1, \ldots, M_k$) achieves the formal functionality of $\mathcal{M}$.

At each step of specification refinement, we need to make sure that the specifications of the components of $\mathcal{M}$ together *cover* the functionality of $\mathcal{M}$, that is, the new specification does not admit any behavior that is invalid with respect to the original specification. The converse is not necessarily true. Since new level specific design constraints are added into the design during design refinement, it is quite possible that the new specification does not admit some behaviors that were acceptable to the original specification. The design intent coverage method presented in Chapter 6 enables the validation engineer to check the validity of each specification refinement step, and to plug the gaps in the new specification.

There are many advantages of this approach:

1. The specification refinement flow is scalable, since we work only with specifications.

2. It helps the validation engineer to prove that the decomposition of the design is correct. Moreover, since this proof works only on the specs, we can do it *before* we invest in writing the RTL of the modules.

3. It facilitates reuse. In future design IPs will come with verification IPs that specify the properties that are guaranteed by the design IP. Such properties can be directly used in the specification refinement proof (through design intent coverage).

4. It systematically allows the development of the formal RTL specs of the unit level modules, ensuring at each step that these specs together cover the micro-architectural specs.

5. The specs at each level act as the coverage goal for the lower levels. This helps us in deciding whether we have written enough properties at a given level.

Specification refinement is actually a formal way of developing a formal verification test plan. The target of the test plan is to cover the micro-architectural specification through local properties over small modules (ones that can be handled by existing FPV tools).

Does specification refinement solve all our problems? In reality, no. This is because specification refinement is easier said than done. Given a micro-architectural property over a design, $\mathcal{M}$, it is conceptually hard to find the set of properties over the components of $\mathcal{M}$ that can together cover the micro-architectural property. There are two issues here:

1. The property specification language may not be expressive enough to specify the set of properties that cover the architectural property. This is rarely the case, since the new property specification languages are very powerful – perhaps more powerful than they need to be.

2. The validation engineer is unable to come up with the desired set of properties. This will be a real problem during the inception of FPV in a company, but will reduce considerably with the experience of the validation engineer in handling complex specifications. The support received from

the intent coverage tool (which points out the gaps) will play a significant role here.

Specification refinement provides a formal mechanism to scale FPV technology by using human ability to decompose functional specifications.

There are many missing pieces in the validation flow consisting of simulation, model checking and specification refinement. We will now look into these pieces.

8.3.1 What's Between Simulation and Model Checking?

The gap between simulation and model checking has been extensively studied in the last decade. This research has led to the development of several hybrid techniques. Broadly, these methods can be classified into two heads:

1. *Formal property verification aided by simulation*

2. *Simulation aided by formal property verification*

There has been a lot of research on each of the above. We present a flavor of some of the interesting ideas that have been productized in some of the recent tools.

FPV Aided by Simulation

The main bottleneck of FPV is in its space complexity. If we can fit the state machine of the implementation in memory, then FPV works well. Otherwise it runs into capacity issues.

This limitation has encouraged many researchers to study the task of creating abstractions of the implementation. Some abstractions preserve the truth of properties in given languages. For example bisimulation equivalence abstractions preserve most temporal languages, and stuttering equivalence abstractions preserve untimed temporal logics without the X operator. These abstractions are *safe* – if our property fails on the abstraction, then it is guaranteed to fail in the implementation, and vice versa.

Safe abstractions do not always give us the desired reduction in space – they are often too big. It is possible to create much smaller abstractions, but they may not preserve the truth of all the properties in the specification. In such cases two things may happen:

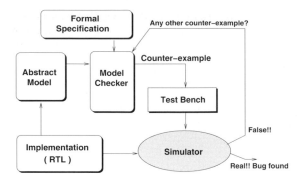

Fig. 8.9. FPV aided by simulation

1. A property evaluates to true in the abstract model, but is actually false in the implementation. This means that a bug escapes detection, and this defeats the whole purpose of *formal* property verification.

2. A property evaluates to false in the abstract model, but is actually true in the implementation. This means that our abstract model produces false counter-examples – ones that do not actually exist in the implementation.

The first is obviously the worse among the two. Therefore, abstractions are made in such a way that the first never happens. In other words, if the abstract model satisfies the specification, then the implementation is guaranteed to satisfy the specification.

How do we verify the soundness of a counter-example? This is where we use simulation (Fig 8.9). We want to verify whether the counter-example trace produced by the model checker from the abstract model is real, that is, whether we can reproduce the counter-example in the implementation. We perform this test by simulating the implementation using the valuations of the input signals in the counter-example trace. If the simulation reproduces the counter-example, then we have indeed found a bug. This method is popularly known as *counter-example guided abstraction refinement* [40].

Simulation Aided by FPV

Typically companies prefer to use FPV before simulation. One common question that is often heard is – *does FPV help in reducing my simulation time in any way?*

The answer is quite intricate. Many companies view FPV as a technology for uncovering difficult corner case bugs. They do not expect FPV to influence simulation effort – thereby, they prefer to run the entire simulation anyway.

The question is therefore heard more often in companies that are in the process of adopting formal verification into their validation flow, and are hesitant to invest in the additional time and manpower. For them, there are two answers to this question.

1. Within its capacity limitations, FPV runs substantially faster than simulation. Therefore the common bugs are uncovered quickly, and the time to verify the fix is also less. This reduces the number of times that we need to repeat the simulation.

2. Test plan coverage by FPV will become a reality, once the test plan structure becomes rigid.

Some of the leading chip design companies use languages having very limited expressibility for specifying their test plan. These specifications are used for automatic generation of the tests (and the test bench). It is possible to check whether these semi-formal specifications are covered by a given set of formal properties. We believe that in future it will be possible for a validation engineer to reduce the number of simulation tests by FPV.

In Chapter 7 we presented a methodology for using formal methods for intelligent test generation. We believe that approaches like this will help us to guide the simulation so that we achieve higher coverage of the interesting behaviors in less time.

8.3.2 What's between Intent Coverage and Model Checking?

High level verification is all about comparing two entities – a *specification* and an *implementation*. Typically there can be many implementations that realize a given specification. The verification goal is to determine whether the given implementation is one among these.

In formal property verification, the specification consists of a set of temporal properties. The power of property specification languages enable us to specify formal properties at any level of the design hierarchy.

What *is* the implementation? FPV tools typically accept RTL code as the implementation. *Is RTL code the only way to define the implementation?*

The answer is, no. RTL code is too low level, and therefore too big. FPV tools are unlikely to be able to handle large designs if the implementation is RTL code.

Design intent coverage is also a form of verification, where both the specification and implementation are sets of formal properties – only the implementation is a more detailed set of properties. In other words, our implementation here is at a more abstract level than RTL code – it is smaller in size and thereby more amenable to formal analysis.

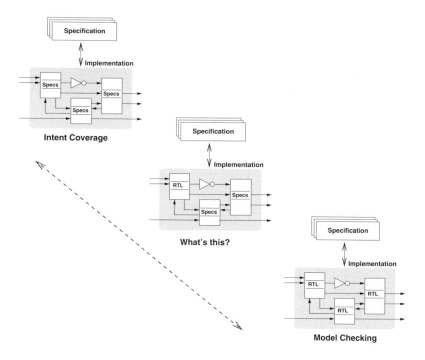

Fig. 8.10. Intent Coverage and Model Checking

If we are given the architecture of a design and the formal specification of the design, then model checking and design intent coverage represent the two ends of a spectrum (Fig 8.10). For model checking we need the RTL code for each block in the architecture. For design intent coverage we need formal property specifications for each block in the architecture. In both cases, the reference is the architectural specification spanning all the blocks.

What lies between design intent coverage and model checking?

In practice validation engineers will have to explore this gap because:

1. Model checking alone will not work due to capacity limitations

2. Design intent coverage alone will not work, since we will always have some modules whose functionality cannot be adequately captured through formal properties.

Therefore in order to *prove* an architectural property over a design, we will have to use a combination of design intent coverage and model checking.

The basic idea is as follows. We need to be able to handle RTL blocks in the design intent coverage proof. *How can we do this?*

1. We may extract the relevant properties from the given RTL blocks, which, along with the RTL properties of the other blocks can be used to establish an intent coverage proof of coverage of the architectural specification.

2. We may find the gap between the architectural specification and the RTL specification and then attempt to prove the properties representing the gap on the RTL components (using model checking).

A recent paper from our group [46] presents more details on these approaches.

It is important to understand where the competetive advantage lies. If we allow too many RTL blocks into the intent coverage approach, then we will end up with the same capacity issues as in model checking. Therefore at the higher levels of the design hierarchy, we should persuade the validation engineers to express the key functionality of the blocks through abstract formal properties. Ideally, at the highest level, only the glue logic used to interconnect the architectural blocks should be in RTL, and all the blocks should be described through properties. As we go down the design hierarchy, we will have smaller modules to deal with – ones that are amenable to model checking and also amenable to the hybrid approach outlined above.

8.3.3 What's between Intent Coverage and Simulation?

We have seen that the intent coverage methodology helps us in setting up a formal verification test plan for the RTL design. We have also studied some of the links between simulation and FPV. *Is there any direct link between intent coverage and simulation?*

We do not yet have a good answer to this question. We do have some preliminary results that may set up a bridge between the two technologies in future. We briefly outline these methods.

1. The specifications of most modules in a design are *assume-guarantee* in nature. In assume-guarantee properties we expect certain behaviors from the DUT under specific assumptions on the inputs. Simulation targets the input scenarios that satisfy the assumptions. Typically a randomized test bench may be viewed as a non-deterministic state machine that generates inputs for the DUT under given constraints. Languages such as SVA allow the designer to specify constraints on the input signals at various states of the test bench – at a given state, the randomizer is constrained by the user specified constraints at that state.

Intent coverage can act as the link between FPV and simulation by constraining the test bench to those scenarios that are not covered through FPV. Recall that the primary goal of intent coverage is to show the validation engineer the gap between the architectural specification and the RTL specification consisting of the local properties of the component modules.

The key technology here is to extract the *assume* part from the assume-guarantee specification representing the coverage gap between the architectural specs and RTL specs. This *assume* part is then mapped into constraints that further constrain the randomized test bench, thereby confining the search for a bug into those parts of the input space that were not covered by FPV. A recent paper from our research group [16] gives more details on this approach.

2. We have presented the intent coverage methodology as a means of comparing specifications at two different layers of the design hierarchy. Intent coverage can also be used between assume-guarantee specifications at a given layer of the design.

For example, if module C receives its inputs from module A and module B, then the *guarantee* part of the specifications of A and B should satisfy the *assume* part of the specification of C. If not, then A and B can potentially drive such inputs into C that do not satisfy its assumptions on its inputs. Therefore, it becomes imperative to verify whether module C satisfies its guarantee also for such inputs. If FPV fails to handle C, then intent coverage can find the offending input scenarios and drive the simulation into those areas.

We believe that several other symbiotic relations between the three key technologies, namely, simulation, FPV and intent coverage, will evolve in the years to come.

8.4 The Integrated Flow

In this section we will attempt to construct an integrated validation flow with the three key technologies and the linkages between them as outlined in the previous section. The focus will be on architectural validation and RTL validation. FPV techniques are also used in the lower levels of the design hierarchy, but a study of those applications is beyond the scope of this book.

8.4.1 Architectural Validation

Architectural validation is all about creating the design intent. Today this is the most critical stage in the design flow in terms of intellectual property. Today most of the traditional optimization goals, such as timing, area, power, testability, are taken up at the architectural level. Some of the leading chip design companies believe that the competitive advantage in their designs are mostly due to the right choices at the architectural level.

Architectural validation has two main challenges:

1. Modeling the architecture and verifying the correctness of the model.

2. Verifying the consistency of the architectural decisions.

Chip design companies use a wide range of frameworks for modeling the architecture. Examples include C, C++, SystemC, System Verilog, SpecC, Esterel, Bluespec Verilog, and SDL. The architectural model is an executable that demonstrates the desired behavior from an abstract level.

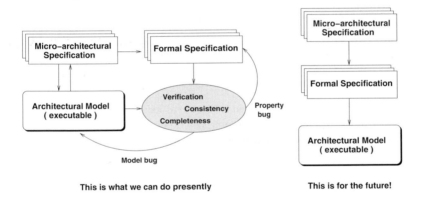

Fig. 8.11. Architectural Validation: *New Flow*

We believe that formal specifications will play an increasingly significant role in architectural validation in the years to come. Fig 8.11 shows two integrated flows in architectural validation – one for the present and one for the future. We explain each of these flows below.

1. Today it is premature to expect that the full architectural specification will be specified formally. Therefore, it is only the key architectural decisions that will be specified through formal assertions. Typical examples include caching and arbitration policies, bus protocol properties, addressing modes, and instruction set architectures.

What will we do at the architectural level with these properties?

a) We will use consistency checking algorithms to verify whether these properties do not conflict with each other. This is an important issue, since inherent inconsistencies in the architectural specification is often hard to detect. We had studied some of these issues in Chapter 4.

b) We will verify these properties on the architectural model. This may be done using static or dynamic methods. For static methods, we need to specify the architectural model in a form from which the abstract states of the model can be easily extracted. This verification is mutually beneficial – if the model does not satisfy one of the properties, then either the property is incorrect and must be rectified, or the architecture is flawed.

At the architecture level, we must use coverage methods to check whether we have written enough properties. Both the mutation based approach and the fault based approach (presented in Chapter 5) may be used at this level. For mutation based approaches we may use the architectural model as the reference – at this level the model has modest size, thereby mutation based coverage is feasible.

2. In the future it should be possible to synthesize the architectural model directly from the formal specification. To make this possible, the formal specification should not only contain properties, but also more expressive formal models. Some of the recent specification languages are beginning to claim partial support towards this direction.

Today the bridge between the architectural level and the implementation is the microarchitectural specs document. In the new flow this will be augmented with the formal architectural properties.

8.4.2 RTL Validation

The existing practice in most companies today is to develop large digital designs top-down and verify them bottom-up. In other words, the design team starts with the architectural specs and develops the design top down. At each step, the designers decompose large functional blocks into smaller modules and define the functionality of the smaller modules. These steps continue until the modules are small enough to be coded directly – we call these *unit* level modules.

Today, validation is done bottom-up. We verify the unit modules first and then proceed bottom-up until we are left with the complete integrated design. This is natural in a simulation based validation flow, since simulation can begin only when we have the complete code for a fragment of the design.

In the integrated flow, we will have a top-down formal verification flow, and a bottom-up simulation-based flow. The key technology in the top-down flow will be intent coverage, and the key technology in the bottom-up flow will be simulation. The two flows will meet at the unit level (or at the level of small blocks), where model checking will play a key role.

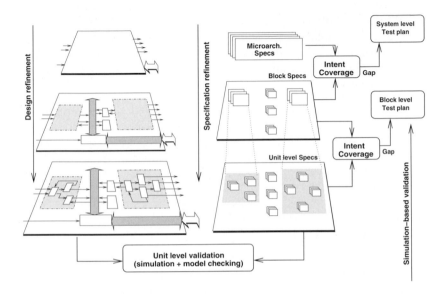

Fig. 8.12. RTL Validation: *New Flow*

Fig 8.12 shows the integrated flow. The design and validation teams will have to work together when the design is developed top-down. Whenever, the design team decomposes the functionality of a block into a set of modules, the validation team will write the formal functional specification of the component modules, and check (using intent coverage) whether it meets the properties of the parent block. This approach will benefit the design team as well – for example, in Chapter 6 we had shown a case where the flaw in the design planning could be detected *before* writing any RTL code.

Therefore at the block level (other than the unit level), the validation team has the following responsibilities:

1. Developing the formal specification of the modules at the next level.

2. Verifying whether the specifications of the modules taken together cover the properties of the block (intent coverage). Typically this check will show that some of the block properties are covered while the rest are not covered.

a) *What shall we do with the properties that are covered?* We will relegate the responsibility of verifying the specifications of the child modules to the next levels, and this will formally guarantee the verification of these properties – this guarantee is by virtue of intent coverage.

b) *What should we do with the block properties that could not be covered by intent coverage?* In the top-down flow we will save these properties. When the bottom-up simulation based validation flow picks up this block for validation we will verify these properties dynamically. At that time, we may also use the techniques presented in Chapter 7 to guide the simulation into the behaviors for which these properties were written.

At the unit level, we may assume that FPV will be attempted before simulation. In the new flow, the top down validation flow will create the properties that are to be checked on each unit module. We will use FPV tools to check these properties and follow it up with simulation. This will mark the start of the bottom-up simulation-based validation flow.

8.5 Sharing the Task

This is the most *delicate* section of this chapter. If we are to adopt the integrated verification flow, then how should we delegate responsibilities? Today several questions are often heard as companies share their experience in adopting formal property verification in their design flows.

• Should micro-acrhitects write formal properties?

• What are the responsibilities of the formal validation team? Writing properties, or running the FPV tools?

• Should designers use FPV tools? Should the sign-off criteria for a unit level designer include the compliance with a set of formal properties?

Since this is a collective responsibility, it is hard for an individual to answer all these questions. It is harder still for an academic (who has none of these responsibilities) to answer these questions.

8.5.1 Architect's Corner

If we adopt the new validation flow, then we will have to write the first formal specification at the architecture level. We will also have to do this *before* implementation starts.

Who will write this specification? Should architects be persuaded to write formal specifications?

Writing formal specifications in a structured and readable form is considered by many as an art, just as developing good architectures is an art. Therefore, there is a difference between *understanding* what is correct and *expressing* what is correct. At the architectural level, we need people dedicated to the art of developing formal property specifications. Let us call these people *formal verification leads*. It is possibly premature to expect microarchitects to become masters of property specification during the inception of the new flow.

Nevertheless, architects will have to participate in the development of the formal architectural specification. They will have to convey to the formal verification leads the exact correctness features of the design architecture. They will have to help in analyzing the completeness of the formal specification by specifying the types of behaviors (faults) that the architecture should not admit.

8.5.2 Design Manager's Corner

The design manager's role in the new validation flow is very crucial. The design manager must *believe* that the new flow benefits her the most. The top down phase of the new validation flow helps in identifying defects in the planning of the design. If these defects are not detected in the top-down phase then they will surface during the bottom-up phase much later in the design cycle. When the defect is detected, the specification of several blocks in the design may have to be changed leading to major changes in the RTL code, and that too for no fault of the RTL designers. In the new approach, we will detect (and rectify) the defect in the top-down phase *before* creating the RTL for that block.

Therefore, it is up to the design manager to establish a protocol between the design team and the validation team. Whenever the functional specification of a design block is decomposed into a set of smaller blocks, the validation team must be called in to develop the formal specification for the smaller blocks and check whether there is any flaw in the decomposition. This process will also help in giving the validation team a head start in developing the simulation test plan for the bottom-up phase of the validation flow.

The intent coverage technology requires a good amount of skill in understanding and working with formal specifications. This task cannot be relegated to designers. Formal verification leads in the validation team must supervise the specification refinement steps, so that we achieve a good coverage of the architectural properties.

8.5.3 Unit Designer's Corner

Who then will use the model checking tools? We believe that this responsibility will ultimately be pushed to the unit designer. Given a set of formal properties on a reasonably small module, FPV works well as an automated methodology. On finding a bug, it produces a counter-example trace (say, as a timing diagram), which is perfectly legible to the designer.

We believe that in future, the sign-off criteria for the design of a unit module should include the check that all given formal properties passed on the design. This will reduce the number of iterations between the design and validation teams.

8.6 Concluding Remarks

It is hard to define *the* roadmap for formal property verification. There are many missing pieces in the roadmap that we have presented in this book. A *roadmap* (by definition) looks into the future – the FPV community will always like to create roadmaps that show the promise of the core FPV technology in design validation.

Over the last five years, we have attempted to create a roadmap for formal property verification that will go beyond the core technology of model checking and will open up the possibility of setting up a unified verification flow that uses formal methods at all levels of the design hierarchy. We have attempted to find the missing technologies in this roadmap and have been able to address some of the major challenges. On the way, we have had the opportunity of finding some formidable partners, including Intel and Synopsys, who have vastly helped us in forming this vision.

There has been several important realizations during the conceptualization of this roadmap.

1. Model checking has hit a complexity barrier which we cannot adequately overcome by adopting engineering ideas in FPV tools.

2. If we can develop large designs by decomposition, then we should also be able to verify large designs by decomposition. If design decomposition can be conceived by human beings, then specification decomposition can also be conceived by human beings. We only need to provide formal methods to verify whether the decomposition of a formal specification is correct.

3. Formal verification is not an alternative to simulation. The main benefits of FPV are at the higher levels of the RTL design hierarchy – to ver-

ify whether the design is being developed in accordance to the design's architectural intent.

4. Verifying the consistency and completeness of formal property specifications will become an increasingly significant problem. We need good tools for this purpose, if FPV is to be adopted by many.

5. Specification guided automatic test generation and simulation will become a reality. We will have to look for bugs in the right places – it will be increasingly infeasible for simulation to look for them everywhere.

All these points out to the fact that FPV technology is poised at an interesting state today. Many chip design companies realize that this technology holds tremendous potential, and yet are unable to seamlessly adopt the technology into their validation flow because of the missing pieces in the integrated validation flow.

These missing pieces offer great opportunities to EDA companies. It is possible to come up with an arsenal of new tools that will scale FPV technology and also enable its integration into a simulation-dominated validation flow.

References

1. Abadi, M., Lamport, L., Conjoining Specifications, In *ACM Transactions of Programming Languages and Systems (TOPLAS)*, Volume 17, Issue 3, pp. 507-535, 1995.
2. Abraham, J.A., Vedula, V.M., and Saab, D.G., Verifying Properties Using Sequential ATPG. In *Proc. of Int. Test Conf. (ITC)*, pp. 194-202, 2002.
3. ACCELLERA.
4. Alur, R., Courcoubetis, C., and Dill, D., Model checking for real time systems. In *Proc. of the 5th Symp. on Logic in Computer Science*, 414-425, 1990.
5. Alur, R., Courcoubetis, C., and Dill, D., Model checking in dense real-time. *Information and Computation*, 104 (1), 2-34, 1993.
6. Alur, R., and Henzinger, T.A., A really temporal logic. *Journal of the ACM*, 41 (1), 181-204, 1994.
7. Alur, R., Henzinger, T.A., Kupferman, O., Alternating-time Temporal Logic. *Journal of the ACM*, 2002.
8. Ammann, P.E., Black, P.E., and Majurski, W., Using Model Checking to Generate Test from Specifications, In *Proc. of 2nd IEEE Int. Conf. on Formal Engineering Methods (ICFEM'98)*, Eds. John Staples, Michael G. Hinchey, and Shaoying Liu, IEEE Computer Society, pp. 46-54, 1998.
9. *ARM AMBA Specification Rev 2.0*, http://www.arm.com
10. Armoni, R., *et. al.* The ForSpec Temporal Logic. In *Proc. of TACAS'2001*, 2001.
11. Armoni, R., *et. al.*, Enhanced Vacuity Detection in Linear Time Logic. In *Proc. of Computer Aided Verification* (CAV), LNCS 2725, 368-380, 2003.
12. Banerjee, A., and Dasgupta, P., The Open Family of Temporal Logics: Annotating Temporal Operators with Input Constraints. *ACM Trans. on Design Automation of Electronic Systems* (TODAES), 2005.
13. Banerjee, A., Pal, B., Das, S., Kumar, A., and Dasgupta, P., Test Generation Games from Formal Specifications, In *Proc. of Design Automation Conference* (DAC'2006), San Francisco, 2006.
14. Basu, P., Das, S., Dasgupta, P., Chakrabarti, P.P., Mohan, C.R., and Fix, L., Formal Verification Coverage: Are the RTL-properties covering the Design's Architectural Intent? In *Proc. of Design Automation and Test in Europe* (DATE), 2004.

15. Basu, P., Das, S., Banerjee, A., Dasgupta, P., Chakrabarti, P.P., Mohan, C.R., Fix, L., and Armoni, R., Design Intent Coverage – A new paradigm for Formal Property Verification. *IEEE Trans. on CAD*, 2006.
16. Basu, P., Das, S., Dasgupta, P., and Chakrabarti, P.P., Discovering the Input Assumptions in Specification Refinement Coverage. In *Proc. of Asia South Pacific Design Automation Conference* (ASPDAC), Yokohama, 2006.
17. Biere, A., Cimatti, A., Clarke, E.M., and Zhu, Y., Symbolic Model Checking without BDDs. In *Lecture notes in Computer Science*, 1579, 193-207, 1999.
18. Biere, A., Cimatti, A., Clarke, E.M., Fujita, M., Zhu, Y., Symbolic Model Checking Using SAT Procedures Instead of BDDs. *Proc. of Design Automation Conference*, 1999.
19. Bergeron, J., Cerny, E., Hunter, A., and Nightingale, A., *Verification Methodology Manual for System Verilog*, Springer Verlag, 2005.
20. Boppana, V., Rajan, S.P., Takayama, K., and Fujita, M., Model Checking Based on Sequential ATPG, In *Proc. of Computer Aided Verification (CAV)*, pp. 418-430, 1999.
21. Bryant, R., Graph-based Algorithms for Boolean-function Manipulation. *IEEE Trans. on Computers*, C-35 (8), 1986.
22. Bryant, R., and Chen, Y.A., Verification of Arithmetic Circuits with Binary Moment Diagrams. *Proc. of Design Automation Conference*, 1995.
23. Buchi, J., Landweber, L., Solving sequential conditions by finite state strategies. *Trans. of the American Mathematical Society*, 138, 295-311, 1969.
24. Budd, T.A., Mutation analysis: Ideas, examples, problems and prospects. *Computer Program Testing*, 129-148, 1981.
25. Burch, J.R., Clarke, E.M., and Long, D.E., Representing circuits more efficiently in symbolic model checking. In *Proc. of Design Automation Conference*, 1991.
26. Burch, J.R., Clarke, E.M., and Long, D.E., Symbolic model checking with partitioned transition relations. In *Proc. of Int. Conf. on VLSI*, 1991.
27. Burch, J.R., Clarke, E.M., McMillan, K.L., Dill, D.L., and Hwang, L.J., Symbolic Model Checking: 10^{20} states and beyond. *Information and Computation*, 98(2), 142-170, 1992.
28. Burch, J.R., Clarke, E.M., Long, D.E., McMillan, K.L., and Dill, D.L., Symbolic Model Checking for Sequential Circuit Verification. *IEEE Trans. on CAD*, 13(4), 401-424, 1994.
29. Callahan, J., Schneider, F., and Easterbrook, S., Automated Software testing using model-checking, In *Proc. of SPIN Workshop*, 1996.
30. Chandy, K.M, Misra, J., Proofs of Networks of Processes, In *IEEE Transactions on Software Engineering*, Volume SE-7, No. 4, pp. 417-426, 1981.
31. Chockler, H., Kupferman, O., Kurshan, R.P., and Vardi, M.Y., A practical approach to coverage in model checking, In *Proc. of Computer Aided Verification (CAV)*, 66-78, 2001.
32. Chockler, H., Kupferman, O., and Vardi, M.Y., Coverage metrics for temporal logic model checking. In *Proc. of TACAS*, LNCS 2031, 528-542, 2001.
33. Chockler, H., and Kupferman, O., Coverage of implementations by simulating specifications. In *Proc. of TCS*, 223, 409-421, 2002.
34. Chockler, H., Kupferman, O., and Vardi, M.Y., Coverage Metrics for Formal Verification. In 12th *Advanced Research Working Conf. on Correct Hardware Design and Verification Methods*, LNCS 2860, 111-125, 2003.

35. Clarke, E.M., Emerson, E.A., and Sistla, A.P., Automatic verification of finite-state concurrent systems using temporal logic specifications, *ACM Transactions on Programming Languages and Systems (TOPLAS)*, 8(2), pp. 244-263, 1986.
36. Clarke, E.M, Filkorn, T., and Jha, S., Exploiting symmetry in temporal logic model checking, In *Proc. of Computer Aided Verification* (CAV), 450-462, 1993.
37. Clarke, E.M., Grumberg, O., and Hamaguchi, K., Another look at LTL model checking, In *Proc. of Workshop on Computer Aided Verification (CAV)*, 1994.
38. Clarke, E.M., Grumberg, O., and Peled, D.A., *Model Checking*, MIT Press, 2000.
39. Clarke, E., German, S., Lu, Y., Veith, H., and Wang, D., Executable Protocol Specification in ESL, In In Proceedings of Formal Methods in Computer-Aided Design (FMCAD), 2000, pp. 197-216.
40. Clarke, E.M., Grumberg, O., Jha, S., Lu, Y., and Veith, H., Counter-example guided abstraction refinement. In *Proc. of Computer Aided Verification* (CAV), 154-169, 2000.
41. Clarke, E.M., Biere, A., Raimi, R., and Zhu, Y., Bounded Model Checking using satisfiability solving. *Formal Methods in System Design*, 19(1), 7-34, 2001.
42. Clarke, E.M., Gupta, A., Kukula, J.H., and Strichman, O., SAT-based abstraction refinement using ILP and machine learning techniques. In *Proc. of Computer Aided Verification* (CAV), 265-279, 2002.
43. Courcoubetis, C., Vardi, M.Y., Wopler, P., and Yannakakis, M., Memory efficient algorithms for the verification of temporal properties. *Formal Methods in System Design*, 1, 275-288, 1992.
44. Das, S., Basu, P., Banerjee, A., Dasgupta, P., Chakrabarti, P.P., Mohan, C.R., Fix, L., Armoni, R., Formal Verification Coverage: Computing the Coverage Gap between Temporal Specifications, In *Proc. of ICCAD*, San Jose, 198-203, 2004.
45. Das, S., Banerjee, A., Basu, P., Dasgupta, P., Chakrabarti, P.P., Mohan, C.R., Fix, L., Formal Methods for Analyzing the Completeness of an Assertion Suite against a High-Level Fault model, In *IEEE International Conference on VLSI Design*, 2005.
46. Das, S., Basu, P., Dasgupta, P., and Chakrabarti, P.P., What lies between Design Intent Coverage and Model Checking? In *Proc. of Design Automation and Test in Europe* (DATE), 2006.
47. Dax, C., and Lange, M., Game Over: The Foci Approach to LTL Satisfiability and Model Checking, Electr. Notes Theor. Comput. Sci. 119(1): 33-49, 2005.
48. Dill, D.L., *Trace Theory for Automatic Hierarchical Verification of Speed-independent Circuits*. ACM Distinguished Dissertations. MIT Press, 1989.
49. eRM – e Reuse Methodology, http://www.verisity.com/products/erm.html
50. Emerson, E.A., and Clarke, E.M., Characterizing correctness properties of parallel programs using fixpoints, In *Automata, Languages and Programming*, LNCS, vol. 85, pp. 169-181, Springer, 1980.
51. Emerson, E.A., and Lei, C.L., Modalities for model checking: Branching time strikes back, In Proceedings of 12th *ACM Symposium on Principles of Programming Languages*, pp. 84-95, 1985.
52. Emerson, E.A., and Halpern, J.Y., Sometimes and Not Never Revisited: On Branching Versus Linear Time, *Journal of ACM*, 33(1), pp. 151-178, 1986.
53. Emerson, E.A., Mok, A.K., Sistla, A.P., and Srinivasan, J., Quantitative temporal reasoning, In *Proc. of Workshop on Computer Aided Verification (CAV)*, 1990.

54. Emerson, E.A., and Sistla, A.P., Symmetry and model checking. *Formal Methods in System Design*, 9, 105-131, 1996.

55. Fernandez, J.C., Jard, C., Jeron, T., and Viho, G., Using on-the-fly verification techniques for the generation of test suites. In *Proc. of Workshop on Computer Aided Verification* (CAV), 348-359, 1996.

56. Fisler, K., and Vardi, M., Bisimulation minimization and symbolic model checking. *Formal Methods in System Design*, 21(1), 39-78, 2002.

57. Fummi, F., Pravadelli, G., Fedeli, A., Rossi, U., and Toto, F., On the Use of a High-Level Fault Model to Check Properties Incompleteness, In *Proc. of Int. Conf. on Formal Methods and Models for Codesign (MEMOCODE)*, pp. 145-152, 2003.

58. Gannon, J., McMullin, P., and Hamlet, R., Data-Abstraction Implementation, Specification, and Testing, *ACM Transactions on Programming Languages and Systems*, 3(3):211-223, 1981.

59. Gargantini, A., and Heitmeyer, C., Using Model Checking to Generate Test from Requirements Specifications; In *Proc. of Joint 7th European Software Engineering Conf. and 7th ACM SIGSOFT Int. Symp. on Foundations of Software Eng. (ESEC/FSE99)*, pp. 146-162, 1999.

60. Gawlick, R., Segala, R., Sogaard-Anderson, J., and Lynch, N.A., Liveness in timed and untimed systems. In *Proc. of Int. Coll. on Automata, Languages and Programming*, LNCS Vol 820, 166-177, 1994.

61. Giunchiglia,E., Narizzano, M., Tacchella, A., System Description: QuBE A System for Deciding Quantified Boolean Formulas Satisfiability. In *Proc. of Intl. Joint Conf. on Automated Reasoning (IJCAR-01)*, 2001.

62. Godefroid, P., and Wolper, P., A partial approach to model checking. In *Proc. of Logic in Computer Science* (LICS), 305-326, 1994.

63. Govindaraju, S.G., and Dill, D.L., Verification by approximate forward and backward reachability. In *Proc. of Int. Conf. on Computer Aided Design* (ICCAD), 366-370, 1998.

64. Grumberg, O., and Long, D.E., Model Checking and Modular Verification, *ACM Transactions on Programming Languages and Systems (TOPLAS)*, vol. 16, pp. 843-872, 1994.

65. Gupta. A.,Casavant. A.E.,Ashar P., and Liu X.G., Property Specific Testbench Generation for Guided Simulation, In *Proc. of Asia-South-Pacific Design Automation Conference (ASP-DAC)*, pp. 524-534, 2002.

66. Henzinger, T., Qadeer, S., and Rajamani, S., Decomposing Refinement Proofs using Assume-Guarantee Reasoning, In Proceedings of *International Conference on Computer-Aided Design (ICCAD)*, IEEE Computer Society Press, pp. 245-252, 2000.

67. Hoskote, Y., Kam, T., Ho, P.H., and Zao, X., Coverage estimation for symbolic model checking, In *Proc. of Design Automation Conference*, 1999.

68. Huang, C.Y., and Cheng, K.T., Assertion Checking by combined word-level ATPG and modular arithmetic constraint-solving techniques, In *Proc. of Design Automation Conference*, pp. 118-123, 2002.

69. Hughes, G.E., and Creswell, M.J., *Introduction to Modal Logic*, Methuen, 1977.

70. IBM CoreConnect, http://www-306.ibm.com/chips/techlib/techlib.nsf/techdocs

71. Kang, H.J., and Park, I.C., SAT-based unbounded model checking. In *Proc. of Design Automation Conference*, 840-843, 2003.

72. Katz, S., Grumberg, O., and Geist, D., Have I written enough properties? - a method of comparison between specification and implementation, In *Proc. of Int. Conf. on Correct Hardware Design and Verification Methods (CHARME)*, 280-297, 1999.

73. Kupferman, O. and Vardi, M.Y., Module Checking, In Proceedings of 8th *International Conference on Computer Aided Verification (CAV)*, LNCS 1102, pp. 75-86, 1996.

74. Kupferman, O., and Vardi. M., Weak Alternating Automata are not that weak. In *5th Israel Symp. on the Theory of Computing Systems*, 1997.

75. Kupferman, O. and Vardi, M.Y., Vacuity Detection in Temporal Model Checking, In Proceedings of 10th *International Conference on Correct Hardware Design and Verification Methods (CHARME)*, LNCS 170, pp. 82-96, Springer-Verlag, 1999.

76. Kupferman, O. and Vardi, M.Y., Vacuity Detection in Temporal Model Checking, *Int. Journal of Software Tools for Technology Transfer (STTT)*, 4(2) pp. 224-233, 2003.

77. Kurshan, R.P., Computer Aided Verification of Coordinating Processes: The Automata Theoretic approach, Princeton University Press, 1994.

78. Lamport, L., "Sometimes" is sometimes "Not Never". In *Proc. of ACM Symp. on Principles of Prog. Lang.*, 174-185, 1980.

79. Lichtenstein, O., and Pnueli, A., Checking that finite state concurrent programs satisfy their linear specification. In *Proc. of ACM Symp. on Principles of Programming Languages*, 97-107, 1985.

80. McMillan, K.L., *Symbolic Model Checking*, Kluwer Academic Publishers, 1993.

81. McMillan, K.L., Applying SAT Methods in Unbounded Symbolic Model Checking. In *Proc. of the 14th Intl. Conf on CAV, LNCS 2404*, Springer Verlag, July 2002.

82. Minato, S., Zero-Suppressed BDDs for Set Manipulation in Combinatorial Problems. *Proc. of Design Automation Conference*, 1993.

83. Moskewicz, M., Madigan, C.F., Zhao, Y., Zhang, L., Malik, S., Chaff: Engineering an Efficient SAT Solver. *Proc. of Design Automation Conference*, 2001.

84. OpenVera Assertions LRM 2.0. http://www.open-vera.com

85. Open Verification Library, http://www.accellera.org/activities/ovl.

86. Peled, D., *Software Reliability Methods*, 2001.

87. Peled, D., All from one, one for all: on model checking using representatives. In *Proc. of Computer Aided Verification* (CAV), 409-423, 1993.

88. Perry, D., and Foster, H., *Applied Formal Verification*, McGraw Hill, 2005.

89. Pnueli, A., The Temporal Logic of Programs. In *Proc. of Foundations of Computer Science*, 46-57, 1977.

90. Pnueli, A., In transition from global to modular temporal reasoning about programs. In *Logics and Models of Concurrent Systems*, Vol 13 of NATO ASI series, Springer Verlag, 1984.

91. Pnueli, A., and Rosner, R., On the Synthesis of a Reactive Module. In *Proc. of the 16th ACM Symposium on the Principles of Prog. Lang.*, 1989.

92. Property Specification Language (PSL). http://www.eda.org/ieee-1850

93. Rabin, M. O., Decidability of second order theories and automata on infinite trees. *Trans. of the American Mathematical Society*, 141, 1-35, 1969.

94. Roy, S., Das, S., Basu, P., Dasgupta, P., and Chakrabarti, P.P., SAT-based solutions for consistency problems in formal property specifications for open systems. In *Proc. of Int. Conf. on Computer Aided Design* (ICCAD), 2005.

248   References

95. Reference Verification Methodology (RVM) for Vera,
    http://www.synopsys.com/products/simulation/pdf/va_vol4_iss1_vera.pdf
96. Sheng, S., Takayama, K., and Hsiao, M.S., Effective safety property checking
    using simulation-based sequential ATPG. In *Proc. of Design Automation Con-
    ference (DAC)*, 813-818, 2002.
97. Shimizu, K., and Dill, D.L., Deriving a simulation input generator and a cov-
    erage metric from a formal specification, In *Proc. of Design Automation Con-
    ference (DAC)*, New Orleans, pp. 801-806, 2002.
98. Sistla, A.P., and Clarke, E.M., The complexity of propositional linear temporal
    logics. *Journal of the ACM*, 32(3), 733-749, 1985.
99. Somenzi, F., *CUDD: CU Decision Diagram Package, Release 2.3.0, User's
    Manual*, Dept. of Electrical and Computer Engineering, University of Colorado,
    Boulder, 1998.
100. Stocks, P., Carrington, D., A Framework for specification-based testing, *IEEE
    Transactions on Software Engineering*, 22(11), pp. 777-793, 1996.
101. Sugar Formal Property Language Reference Manual.
    http://www.haifa.il.ibm.com/projects/verification/sugar/
102. System Verilog. http://www.systemverilog.org/
103. Tasiran, S., and Keutzer, K., Coverage metrics for functional validation of
    hardware designs. *IEEE Design and Test of Computers*, 18(4), 36-45, 2001.
104. Tasiran, S., Yuan, Y., Batson, B., Using a formal specification and a model
    checker to monitor and direct simulation, In *Proc. of Design Automation Con-
    ference (DAC)*, pp. 356-361, 2003.
105. Vardi, M.Y., and Wolper, P., An automata-theoretic approach to automatic
    program verification. In *Proc. of Logic in Computer Science* (LICS), 1986.
106. Vardi, M., Verification of Open Systems, In *Proc. of Int. Conf. on Foundations
    of Software Technology and Theoretical Computer Science (FST&TCS)*, pp.
    250-266, 1997.
107. Wang, C., Li, B., Jin, H., Hachtel, G.D., and Somenzi, F., Improving ariadneys
    bundle by following multiple threads in abstraction refinement. In *Proc. of Int.
    Conf. on Computer Aided Design* (ICCAD), 408-415, 2003.
108. Zhang, L., and Malik, S., Conflict Driven Learning in a Quantified Boolean
    Satisfiability Solver. In *Proc. of the Intl. Conf. on CAD*, November 2002.
109. Zhu, H., Hall, P.V., and May, J.R., Software unit test coverage and adequacy.
    ACM Computing Surveys, 29(4), 366-427, 1997.

Index

abstractions, 13
  cone-of-influence, 76
  simulation relations, 77
    bisimulation, 77
    stuttering, 77
Accellera, 31
alternating automata, 117
  very weak alternating automata, 62,
    69, 117, 118
  weak alternating automata, 62
AMBA Bus, 195
approximate coverage, 150
architectural intent, 157
assertion IP, 47
  MyBus protocol, 50
  coding styles, 50
    event-based coding, 53
    factored state-based coding, 55
    state-based coding, 54
  development steps, 49
assertions as coverage points, 203
assume-guarantee verification, 13, 74

binary decision diagram (BDD), 13, 73,
    78
  characteristic functions, 85
  definition of, 79
  equivalence checking, 80
  packages, 83
  reachability using, 84
  reduction steps, 80
  representing state machines, 83
  variable ordering, 80

complexity of, 81
  dynamic reordering, 81
bounded model checking (BMC), 13, 96

channels, 199
characteristic functions, 85
checker automata, 61
compositional verification, 74
consistency, 5, 101–128
consistency checking
  as a game, 115
  methods, 115
consistency issues
  coding errors, 102
  logical errors, 102
  open versus closed, 106
constrained random test, 196
coverage, 6, 129–155
coverage driven test generation, 202
coverage metrics
  fault-based coverage
    coverage strategy, 141
  for simulation, 131–132
    assertion coverage, 132
    circuit coverage, 131
    code coverage, 131
    FSM coverage, 132
    latch coverage, 131
    mutation coverage, 132
    toggle coverage, 131
  FPV coverage, 132–155
    circuit coverage, 136
    code coverage, 136

complexity of, 134
falsity coverage, 134
fault-based coverage, 138–155
FSM coverage, 135
incompleteness of, 133, 138
mutation coverage, 132
path coverage, 135
state coverage, 135
transition coverage, 135
vacuity coverage, 134
functional, 130
structural, 130
structural versus functional, 136
coverage points, 195
coverage-driven test generation, 197

design intent
architectural intent, 157
design intent coverage, 74, 157–194
coverage algorithm, 174
coverage definition, 168
coverage gap, 171
architectural, 173
RTL, 171
detecting architectural flaws, 166
invariant properties, 184
primary coverage question, 169
problem definition, 167
term distribution, 178
directed tests, 195
dynamic property verification, 9, 64
algorithm, 64

e Reuse Methodology (eRM), 48, 197

fault coverage, 139
on inputs, 140, 145
strong coverage, 145
weak coverage, 147
on non-inputs, 140, 143
fault models, 139
counter-example faults, 154
multiple stuck-at faults, 153
short faults, 154
single stuck-at faults, 140
full-x tree, 114

games
strategy, 114

with the environment, 114

layered test environment, 197
benefits of, 199
channels, 199
command layer, 198
functional layer, 198
limitations of, 201
scenario layer, 199
signal layer, 198
test layer, 199
LTL-Covanalyzer tool, 148

model checking, 11
automata theoretic style, 69
on-the-fly, 69, 98
tableau style, 70
word-level model checking, 98
approximate, 13
approximate methods, 74
assume-guarantee, 74
ATPG style, 98
BDD-based methods, 78
complexity of, 12
compositional approach, 74
CTL model checking, 90
FSM extraction, 72
LTL model checking
fair runs, 71
checker construction, 68
main steps, 67
tableau construction, 70
SAT based methods, 92
bounded model checking (BMC), 96
detecting loops, 95
reachability, 93
unfolding properties, 96
state explosion in, 73
symbolic LTL model checking, 91
symbolic model checking, 78
under *assume* constraints, 76

Open Verification Library, 3, 31

Priority cache access, 160, 189
Property Specification Language, 3, 31

quantified Boolean formula (QBF), 117
QuBE QBF solver, 125

randomized test generation, 196
realizability, 105–109, 143, 207, 214
    checking algorithm, 121–123
    in FPV coverage, 143, 144
    in test generation, 207
    QBF representation, 122
receptiveness, 109–113, 207
    checking algorithm, 123–125
    QBF formulation, 125
Reference Verification Methodology
        (RVM), 48, 197
run tree of VWAA, 118
    propositional formula for, 118

SAT, 74
    bounded model checking, 96
    definition of, 92
    model checking, 92
    reachability by, 93
    unfolding properties into, 96
satisfiability, 103–104
    checking algorithm, 119–121
scoreboard directed coverage, 202
simulation coverage metrics, 131–132
SpecChecker tool, 125
specification refinement, 15, 217, 227
    coverage, 16
    reuse, 16
SpecMatcher tool, 188
symbolic model checking, 86
    backward reachability, 89
    CTL model checking, 90
    forward reachability, 89
    LTL model checking, 91
symbolic reachability, 84
symbolic state traversal, 86
System Verilog, 32, 197
System Verilog Assertions, 3, 31
    assert properties, 34, 46
    assume properties, 45, 46
    cover properties, 47
    disable iff clause, 43
    bind, 34, 36
    implication operators, 43
    interface, 34
    local variables, 41
    properties, 42
    sequence expressions, 33, 37
    sequence operations, 38, 40
        AND, 40
        INTERSECT, 40

OR, 40
    consecutive repetition, 38
    goto repetition, 39
    non-consecutive repetition, 39

temporal depth, 117
temporal logics
    Computation Tree Logic (CTL), 27
    Linear Temporal Logic (LTL), 3, 26
    LTL versus CTL, 29
    real time, 30
    rewriting rules, 65, 118
temporal operators, 3, 22
    always, 24
    bounded until, 30
    duality between, 24
    eventually, 24
    future, 3
    global, 3
    next time, 3, 23
    real time, 30
    semantics of, 24
    until, 4, 23
temporal properties
    strength of, 151, 171
    strengthening, 152
    weakening, 182
        by substitution, 182
        by variable weakening, 183
    X-depth, 175
    X-guarded, 175, 208
    X-pushed, 175, 208
temporal worlds, 21
test generation games, 195–215
    algorithms, 210
    dynamic monitoring, 208
    online test generation, 209
    realizability, 207
    role of environment, 204
    vacuity, 205
test plan, 195

unfolding properties, 117

vacuity, 105, 150, 205
vacuous input vector, 209
verification IP, 48
    test environment, 48
Verification Methodology (VM), 48, 197

zchaff SAT solver, 125